Genetics—Principles and Perspectives: a series of texts

General Editors: Dr. K. R. Lewis, Professor Bernard John

Volumes already published

1	**The Differentiation of Cells**	Norman Maclean
2	**The Genetics of Recombination**	D. G. Catcheside, F.R.S.
3	**Chromosome Genetics**	H. Rees, F.R.S. and R. N. Jones

Chromosome Genetics

H. Rees, D.Sc., F.R.S.,
R. N. Jones, Ph.D.

Department of Agricultural Botany, University College of Wales,
Aberystwyth.

Edward Arnold

First published 1977
by Edward Arnold (Publishers) Limited
25 Hill Street, London W1X 8LL

Boards edition ISBN: 0 7131 2626 4
Paper edition ISBN: 0 7131 2627 2

Printed in Great Britain by
William Clowes & Sons, Limited,
London, Beccles and Colchester

Preface

The aim of this book is to present a brief but comprehensive account of the structure and organization of chromosomes, their behaviour during cell division, their variation within and among species, and to explain how the chromosome mechanism regulates the distribution of genetic information during growth and reproduction. During recent years many new facts have come to light about the molecular composition and fine structure of the chromosomes, but we are still a very long way from appreciating what these facts imply in terms of genetics and genetic systems. We have attempted to relate the implications of new knowledge and discoveries at the molecular level to established concepts of the chromosome as a whole.

In a small book of this kind we have been obliged to be selective. Some subjects, not extensively covered in other textbooks, are however presented at some length, *B*-chromosomes for example.

The book is intended for students with an elementary knowledge of genetics. We hope the facts and principles will be of value not only to cytologists and geneticists but to biologists generally.

We are grateful to our colleagues Dr. A. Durrant, Dr. G. M. Evans and Dr. R. K. J. Narayan for helpful comments on the manuscript. We also thank Miss Elizabeth Phillips for typing the script and Mr. E. Wintle for developing and printing some of the photographs. Finally, we thank those who have kindly allowed us to reproduce figures and tables from their published work.

Contents

1

The physical basis of heredity

1.1 Information

Genetics is concerned with the nature and utilization of the heritable information which controls the development of living organisms and with the distribution of this information during growth and reproduction. The information is located mainly within the chromosomes which, in most organisms, are organized within the cell nucleus. The location of heritable information within the chromosomes is directly established by two kinds of experiment: reciprocal grafts and reciprocal crosses.

Reciprocal grafts

Algae of the genus *Acetabularia* are unicellular with a single nucleus in the rhizoid of the filamentous cell. The shape of the cap at the top of the filament varies between species. In *A. mediterranea* the cap is round and smooth edged, in *A. crenulata* it is frilled. When the filament of *A. mediterranea* is grafted on to the nucleated rhizoid of *A. crenulata* the new cap which develops is of the *crenulata* type, and vice versa (Fig. 1.1). Information about cap development is evidently located in the chromosomes, which make up the nucleus, and not in the cytoplasm.

More sophisticated grafts can be achieved by nuclear transplantation (Danielli, 1958). By microdissection it is possible in *Amoeba* to replace the nucleus of one cell by that of another. When the nuclei from one strain are displaced by those of another observations on the nucleus/cytoplasm 'hybrids' and their progenies show some characters to be controlled by nuclear genes, others by genes located in the cytoplasm. Equally important the experiments show instances of interaction between nuclear and cytoplasmic determinants.

Reciprocal crosses

The contribution of material to the zygote by the female gamete in higher plants and animals is vastly greater than by the male gamete. The human egg has a diameter of 140 µm and weighs about 0.0015 mg. The sperm head by comparison has a length of only 4 µm and weighs 0.000 000 005 mg. The mass of the latter is made up almost entirely of the chromosomes. In addition to chromosomes the egg contains a

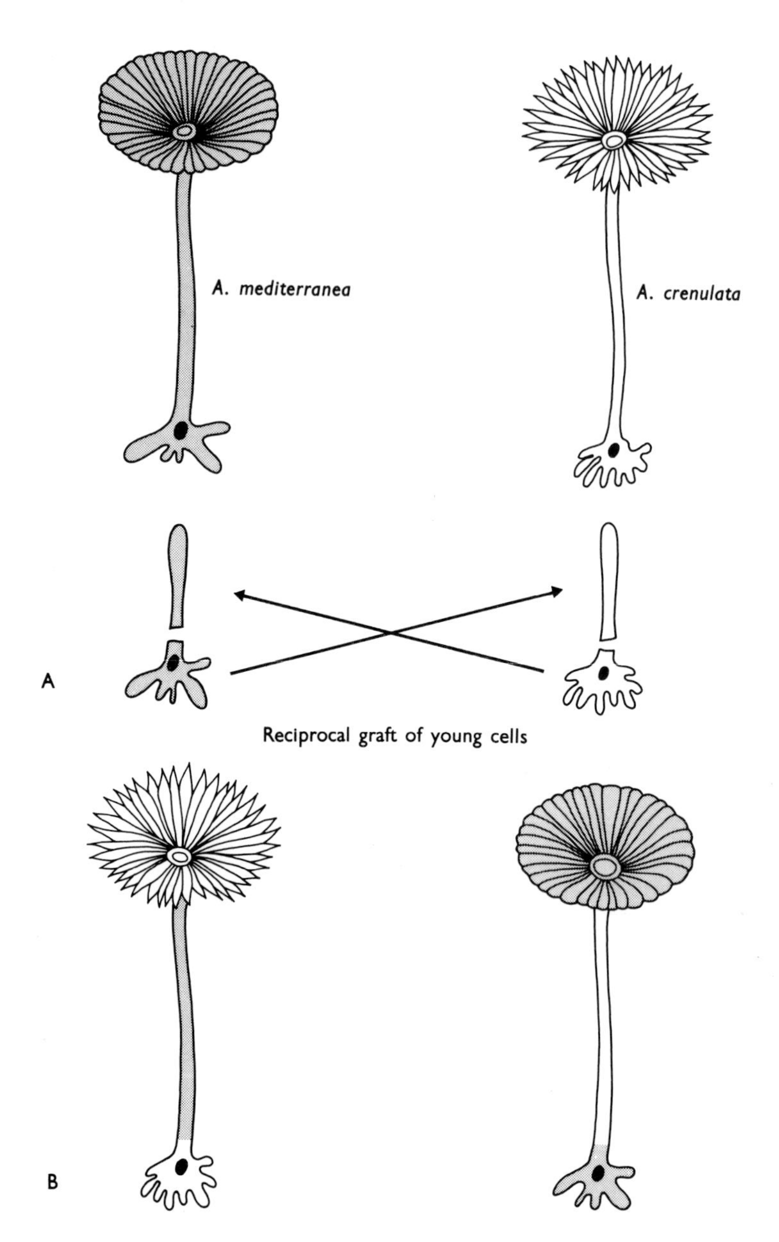

Fig. 1.1 (A) Reciprocal grafts between young cells of two *Acetabularia* species combining, in the main, the cytoplasm of one species of *Acetabularia* and the nucleus of another. (B) The adults display cap phenotypes determined by the nucleus, not the cytoplasm.

massive quantity of extranuclear, cytoplasmic material. A comparable situation applies in higher plants. Yet for the vast majority of characters the results of reciprocal crosses are identical. When Mendel crossed Tall peas with Dwarf, the F_1 and F_2 progenies were similar whether Tall plants were used as male or female parents. The inference is clear, namely that the genetic determinants are located in the chromosomes within the nucleus, not in the cytoplasm. Yet, as with the exceptions in Danielli's grafts in *Amoeba*, the results of reciprocal crosses are sometimes different. Control over certain aspects of the phenotype may be exercised by cytoplasmic 'genes' and by their interaction with the genes in the nucleus.

That the genetic information resides mainly within the chromosomes was established in the 1920s, and in the 1930s the Chromosome Theory of Heredity was elaborated, mainly on the basis of cytological and linkage analyses in *Drosophila* and maize. It was confirmed that the number of linkage groups corresponded with the basic chromosome number; that linkage maps, as one might suspect from the thread-like organization of chromosomes, were linear; and that structural rearrangements within or between chromosomes were accompanied by changes in linkage relations. This information, however, gave no indication of how or where the information was carried by the chromosomes which, in higher plants and animals, are organelles of considerable complexity in both chemical and physical terms. Certain key experiments served to locate the information precisely within the nucleic acid component of the chromosomes.

Virus fractionation and reconstitution

In the small plant viruses, such as Tobacco Mosaic Virus (TMV), individual particles comprise two components, a protein sheath embracing a single strand of ribonucleic acid (RNA). Together they make up a cylinder 300 nm in length and 15 nm in diameter. The disease symptoms exhibited by plants infected by virus are direct consequences of the physiological activities of these virus particles. Symptoms caused by different strains of the virus may differ sharply from one another. The information which determines the differing properties of different strains may be located unambiguously in the nucleic acid component of each particle. The experimental evidence is presented in Fig. 1.2. It will be observed that the nucleic acid component alone is sufficient to induce symptoms of virus disease; also that virus particles reconstituted from two different strains cause disease symptoms characteristic of the strain from which the nucleic acid is derived, and do not show symptoms of the strain which contributed the protein. Following replication within the host, the protein component of new virus particles is identical to that in the strain from which the nucleic acid was extracted. Evidently the nucleic acid also carries the information required for the assembly of its protein envelope.

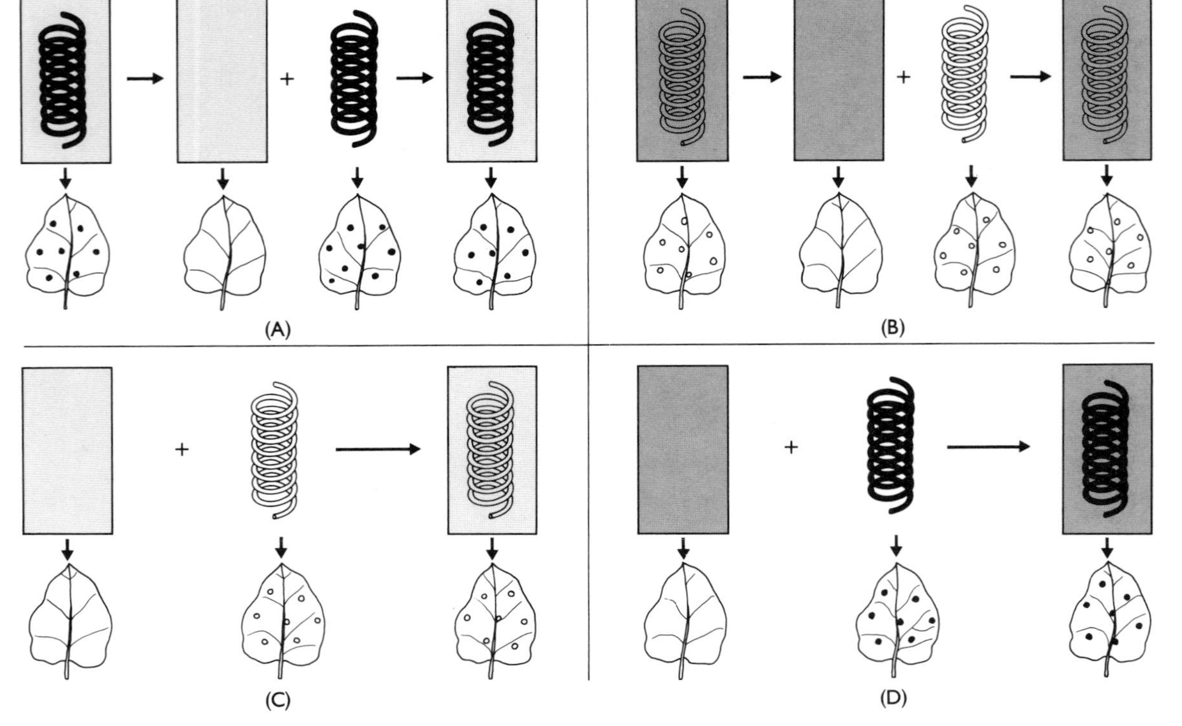

Fig. 1.2 Evidence to show that the nucleic acid (RNA) of the Tobacco Mosaic Virus carries the genetic information. In (A) the virus particle is dissociated into its component parts, namely protein and RNA. The RNA is infectious, the protein is not. Virus particles re-associated from protein and RNA are also infectious. Viruses recovered from leaves infected by RNA alone or re-associated virus particles are identical with the original. (B) shows the same result with another strain. (C) and (D) show that infection by protein/RNA 'hybrid' particles gives symptoms characteristic of

In emphasizing that the genetic information resides in the nucleic acid within the particle this is not to say that the protein may not participate in its *expression*. The ultimate source of information is, however, unquestionably the nucleic acid, the viral RNA. The nucleic acid in TMV is not, of course, organized within a chromosome, unless we regard each individual as a chromosome in itself! Experiments in other organisms where the nucleic acid determinants are embodied within chromosomes serve equally convincingly to locate the information in the nucleic acid of these chromosomes.

Transformation

Griffith (1928) showed that when the debris of *Pneumococcus* cells killed by heat was added to a culture of living *Pneumococcus* of a different strain a small proportion of the latter underwent a permanent change such that they came to resemble in certain respects the strain from which the dead material was derived. For example, debris obtained from a strain in which the cells are highly virulent and coated with a polysaccharide capsule of immunological type III transforms cells of a non-virulent, unencapsulated strain to virulent cells encapsulated in precisely the same way as the transforming strain. The *transformation* is of a permanent, heritable nature as can be established by repeatedly infecting and re-infecting mice with the transformed strain. In 1944, Avery, McLeod and McCarthy showed that the active ingredient of transformation is the nucleic acid, in this case DNA, deoxyribonucleic acid. Purified DNA extracts, and no other single component, effects the transformation.

Bacteriophage infection

Viruses which infect and penetrate bacteria, such as the T4 bacteriophage of *Escherichia coli*, are, relative to other viruses, complex in construction. Even so they are simple in chemical composition, consisting of 40% DNA packed tightly in the head and 60% protein. Within the host the bacteriophage multiplies, utilizing the raw materials of the cell. Hershey and Chase (1952) showed that the information which enables the virus to multiply and develop within the host is vested in the DNA component of the head. They labelled the DNA of the phage with radioactive phosphorus (^{32}P) and the protein with radioactive sulphur (^{35}S). Progenies of these phages produced within the bacteria are heavily labelled with ^{32}P but have incorporated little or none of the ^{35}S. Complementary evidence showing that only the DNA of the phage enters the bacterial cells and that only the DNA is endowed with the genetic information comes from the studies of Hershey (1955). The heads of bacteriophages subjected to osmotic shocks burst to release the DNA leaving empty protein 'ghosts'. These are capable of attaching themselves to the surface of a bacterium but there is no penetration and replication, from which we infer that the DNA alone embodies the information controlling these events.

Table 1.1 Chromosome organization in prokaryotes*

Prokaryote	Nucleic acid	Chromosome configuration†	Chromosome length (μm)	Number of base pairs (or bases)
VIROIDS	1-RNA	L	<0.1	<200
BACTERIAL VIRUSES				
T2, T4	2-DNA	L	54.0	162 000
T5	2-DNA	L	39.0	117 000
λ	2-DNA	L	17.3	51 900
P22, P2	2-DNA	L	13.7	40 500
T7	2-DNA	L	12.5	37 500
ϕ6	2-RNA	L(M)	4.7	14 100
ϕX174	1-DNA	O	1.8	5 400
R17, F2, MS2	1-RNA	L	1.0	3 000
PLANT VIRUSES				
Wound tumour	2-RNA	L	7.0	21 000
Brome mosaic	1-RNA	L(M)	3.0	9 000
Potato virus-X	1-RNA	L	2.9	8 670
Cauliflower mosaic	2-DNA	O	2.7	8 100
Tobacco mosaic	1-RNA	L	2.1	6 300
Turnip yellow mosaic	1-RNA	L	2.0	6 000
ANIMAL VIRUSES				
Fowlpox virus	2-DNA	L?	100.0	300 000
Vaccinia	2-DNA	L	80.0	240 000
Herpes	2-DNA	L	53.0	159 000
Reovirus	2-RNA	L	11.3	34 000
Foot and mouth	1-RNA	L	3.4	10 300
Polio	1-RNA	L	2.6	7 800
Polyoma	2-DNA	O	1.5	4 500
BACTERIA				
Pseudomonas aeruginosa	2-DNA	O?	3480.0	10 440 000
Streptomyces coelicolor	2-DNA	O?	2600.0	7 800 000
Bacillus subtilis	2-DNA	O	1350.0	4 050 000
Escherichia coli	2-DNA	O	1333.0	4 000 000
Neisseria gonorrhoea	2-DNA	O	640.0	1 920 000
Acholeplasma laidawii	2-DNA	O	507.0	1 520 000
Haemophilus influenzae	2-DNA	O	505.0	1 515 000
Mycoplasma hominis	2-DNA	O	253.0	760 000

* Estimates of chromosome length and base pair number have been made using the following approximate mass-length equalities of DNA:
3000 base pairs ≡ 1 μm ≡ M.W.2×10^6.

† L = Linear chromosome, O = Circular chromosome, (M) = Genome apparently divided into several pieces though this may be a consequence of breakage during preparation.

It is only in the smaller viruses that RNA acts as the repository of genetic information. In all other micro-organisms the role is fulfilled by DNA (Table 1.1). In many of the RNA viruses the information is embodied in a single RNA strand. A complementary strand functions only as a template during multiplication of the virus. The situation is similar in viruses which carry a single strand of DNA. Where, as is most often the case, the DNA takes the form of a double helix, only one of the two strands at any one segment (the plus strand) carries base sequences transcribed to messenger and, subsequently, to protein. One means by which this was established was to render the normally double-stranded DNA of the bacteriophage SP8 single stranded, by heating in the presence of messenger RNA (mRNA). At any one segment the mRNA forms duplexes with one only of the DNA strands. It is worth emphasizing that the same strand may be plus at one locus, minus at another.

Higher organisms

The chromosomes of higher organisms all contain DNA and it was reasonable in the light of evidence from bacteriophage and other micro-organisms to conclude that this DNA was the bearer of genetic information. In addition to DNA however the chromosomes contain protein, RNA and traces of other compounds. This complex chromosome composition and organization is characteristic of eukaryotes as distinct from the prokaryotes, in which the chromosomes are naked DNA helices (double in *E. coli*, single in bacteriophage ϕX174) or RNA helices (single in TMV, double in the animal virus, Reovirus). The view that information resides in the DNA fraction is supported by the fact that the DNA content per chromosome is constant whereas other components, the protein in particular, vary in amount (Table 1.2) and quality (Table 1.3). This observation does not, however, provide incontrovertible evidence that the chromosomal DNA carries the genetic information because one could argue that a fraction of the protein is of consistent quality and quantity. More recently, positive proof has been forthcoming in respect of many different kinds of information, i.e. of the information embodied in different genes within the chromosome. One kind of proof comes from labelling the RNA transcribed from the gene with a radioisotope, usually tritium ^{3}H, and hybridizing this RNA to the DNA *in situ* at the locus of the gene on the chromosomes. RNA/DNA hybridization is achieved under conditions which render the chromosomal DNA single stranded, as in the presence of strong alkali. In *Drosophila melanogaster*, for example, the 5*S* (*S* = Svedberg units) fraction of ribosomal RNA, labelled with tritium, hybridizes with the DNA at the 56F band in chromosome 2, the location of a cluster of genes coding for the 5*S* RNA fraction. There are also clusters of genes coding for 18*S* and 28*S* RNA at the nucleolus organizers in both *X* and *Y* chromosomes. In the wild type fly the 18*S* and 28*S* ribosomal cistrons are each reiterated about 130 times per haploid genome (Wimber and Steffensen, 1973). In these instances the information along with the base sequences making up the genes which specify the information are repetitive. Other base sequences within the chromosomes of higher organisms, as we shall see later, are repeated to a far greater degree. Many sequences and the genes they represent are in contrast unique within the chromosome set or genome.

Another sensitive method for tracing the source of information within the chromosomal DNA, including that which emanates from non-repetitive, i.e. single-copy genes, has been employed by Bishop and Freeman (1973). The messenger RNA coding for haemoglobin in blood cells of the duck was extracted and from this a complementary DNA sequence, cDNA, prepared using the enzyme reverse transcriptase. The labelled cDNA was then mixed with single stranded unlabelled duck DNA. The cDNA annealed with the complementary sequences to produce double stranded segments. The results establish not only that the messenger RNA is transcribed by the DNA extract but, on the basis of the extent of hybridization, that the messenger is made up of sequences transcribed at three different loci. Since the haemoglobin includes two kinds of polypeptide chains, α and β, produced by the action of different genes, one of the loci must transcribe for α and the other two for β, or vice versa.

Table 1.2 Constancy of DNA and quantitative variation in other chromosomal components, from different tissues of the fowl. [Data of *Mirsky, A. E. and Osawa, S. (1961) *The Cell*, Vol. II, Academic Press Inc.; †Seligy, V. and Miyagi, M. (1969) *Expl. Cell Res.*, **58**, 27–34]

Component	Tissue		
	Red blood cells	Kidney	Liver
* DNA per nucleus (pg)	2.58	2.28	2.65
† Protein/DNA ratio in isolated chromatin	1.18	1.76	2.61
† RNA/DNA ratio in isolated chromatin	0.003	0.007	0.025

Table 1.3 Qualitative differences in chromosomal histone proteins during early development of Japanese newts. Lewin's nomenclature in brackets. [From Asao, T. (1969) *Expl. Cell Res.*, **58**, 243–52]

Stage of embryo	Histone fractions (ng per 10^4 nuclei)		
	FI (H1, H5)	FII (H2A, H2B)	FIII (H3, H4)
Early gastrula	0	0	58
Middle gastrula	28	47	50
Late gastrula	117	192	114
Neurula	218	409	114
Tail bud	256	526	119

1.2 Eukaryote chromosomes

The chromosomes of eukaryotes, as we have indicated earlier, are complex in structure and composition. While it is clear that the DNA fraction carries the genetic information, the way in which the DNA and other fractions are organized and arranged is by no means clear. It will be useful in the first instance therefore to consider the general morphology of the chromosomes and subsequently, to determine as far as possible how the morphology relates to the separate components and to their organization.

Morphology

Under the light microscope, details of the morphology of the chromosomes are in most species manifested only when the chromosomes are contracted during nuclear division. At interphase they are extended to such a degree that individual members of the complement are indistinguishable. At mitosis the contraction, through coiling and supercoiling, is at a maximum at metaphase. Treatment with colchicine or other spindle inhibitors, increases the contraction and also facilitates the spreading of the chromosomes. From such metaphases it is clear that the chromosomes vary enormously in number and in size among eukaryotic species but remain remarkably consistent within species. A longitudinal differentiation in the structure of individual chromosomes is equally obvious. The pattern of differentiation is important for purposes not only of identity but also of function. The 'markers' of identity and function are as follows.

(a) *Primary and secondary constrictions*

The primary constriction is called the centromere, a less densely coiled and thereby slender segment near which the chromatids are held together during metaphase. The centromere is essential for movement and *acentric* fragments, which lack a centromere, are therefore incapable of movement on the spindle because spindle fibres only attach to centromeres. The chromosomes of the vast majority of species contain one centromere. In a few species of plants, such as *Luzula campestris*, and

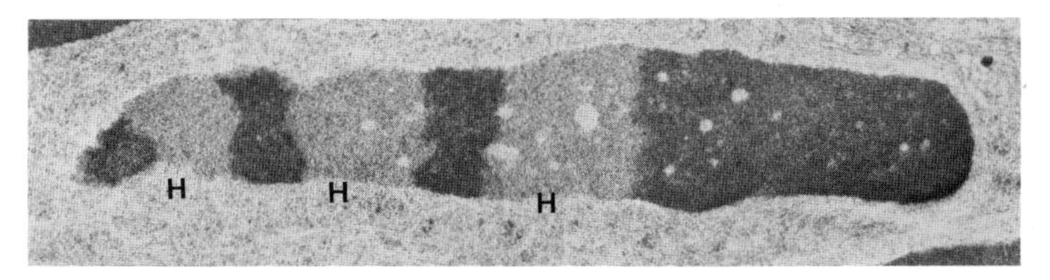

Fig. 1.3 Electron micrograph of one arm of a metaphase chromosome of *Fritillaria lanceolata* showing three heterochromatic (H) segments which are distinguishable by their reduced opacity to electrons. ×20 000. [Courtesy of L. F. LaCour]

animals (e.g. *Pseudococcus obscurus*) the centromeres are *diffuse* and the capacity for mobility on the spindle is spread throughout most of the chromosome. Chromosome fragments in these cases are capable of movement by attachment to spindle fibres.

Secondary constrictions mark the locations at which the nucleoli are assembled. Both primary and secondary constrictions are consistent in location. They reflect an underlying variation in the molecular organization of the chromosomal DNA. They testify also to an association between the coiling status of chromosome segments and their functional activity.

(b) *Heterochromatin*

At interphase in many species, a specific and consistent fraction of the chromosome material, the heterochromatin, is out of phase in being highly contracted when the bulk of the chromosomes, the *euchromatin*, is extended and uncoiled. These same segments are more densely coiled also at early prophase both of mitosis and meiosis. Following cold treatment in many species, the heterochromatin is also revealed at metaphase, being uncoiled and less densely stained than the euchromatin. In *Fritillaria lanceolata* even at normal temperatures the heterochromatin is sharply distinguishable from euchromatin under the electron microscope (Fig. 1.3) by virtue of its reduced opacity to electrons. However, in other species, e.g. *Scilla sibirica*, no such distinction is possible in metaphase chromosomes. The consistency in the distribution of heterochromatin among chromosomes, to which we have referred, is not invariable. Some chromosomes or segments of chromosomes are heterochromatic in some cell environments and not in others. The most familiar example of this kind of heterochromatin is found in the *X*-chromosomes of female mammals. One of the two *X*-chromosomes is heterochromatic in each cell. The heterochromatinization is a random event with respect to the two *X*-chromosomes such that the *X* from the maternal parent is heterochromatic in about half the cells, the *X* from the paternal parent in the other half. This class of heterochromatin is *facultative* as distinct from the *constitutive*.

We have referred above to an association between genetic activity and chromosome coiling. Nowhere is this more apparent than with the heterochromatinization of the mammalian *X*-chromosome (Lyon, 1963). The genes on the heterochromatic *X* are 'switched off'. Because one *X* is heterochromatic in half the cells and its homologue is heterochromatic in the other half, in a genic heterozygote one allele of any *X*-chromosome gene is active in half the cells, the other allele in the remainder. For example, in the female mouse heterozygous for the alleles + and *Mo* which determine dark and light coat colour respectively, the mouse presents a mosaic of dark and pale patches (Fig. 1.4).

Constitutive heterochromatin may be distinguished from the facultative type by staining with fluorescent dyes such as quinacrine mustard and viewing under ultraviolet light. Constitutive heterochromatin

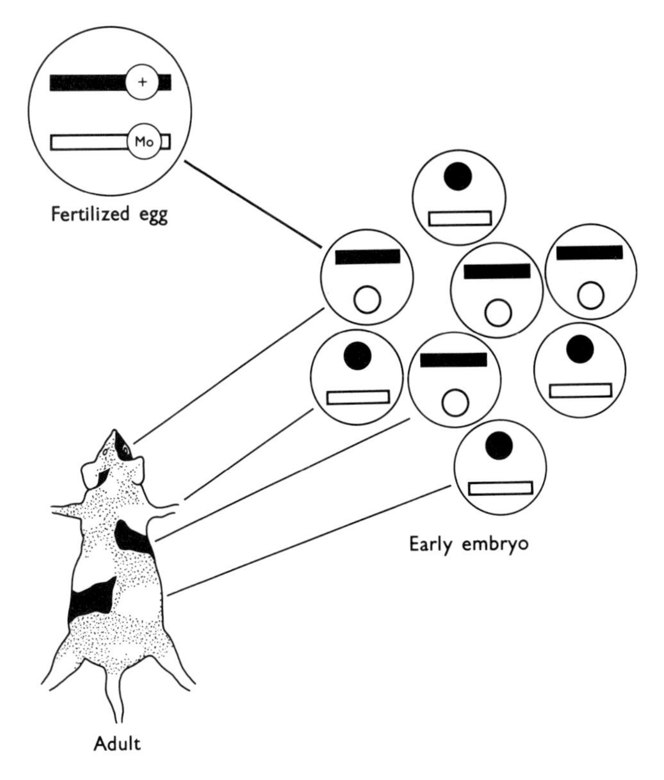

Fig. 1.4 X chromosome inactivation in the female mammal. The inactive X is shown as a circle and, the active one as a rod. [From Lyon, M. F. (1963) *Genet. Res.* **4**, 93–103]

fluoresces, the facultative heterochromatin does not. Staining with Giemsa dye, following treatment with barium hydroxide and subsequently SSC (a mixture of sodium chloride and sodium citrate), is another way of distinguishing the constitutive heterochromatin, which stains violet. Modifications of this method bring to light other segmental differences along the chromosomes: G-bands as distinct from the C-bands which mark the constitutive heterochromatin (Fig. 1.5). What the G-bands specify in terms of organization or function is not clear.

(c) *Puffs, bands and loops*

In the salivary glands of the diptera the chromosomes are multiplied side by side (up to 1000 times) such that the lengthwise variation in morphology becomes amplified to render unusually distinctly the elaborate differentiation of chromosome segments in respect of density of coiling and of staining. The heterochromatin in the salivary gland nuclei of *Drosophila* species, for example, shows up as lightly stained diffuse

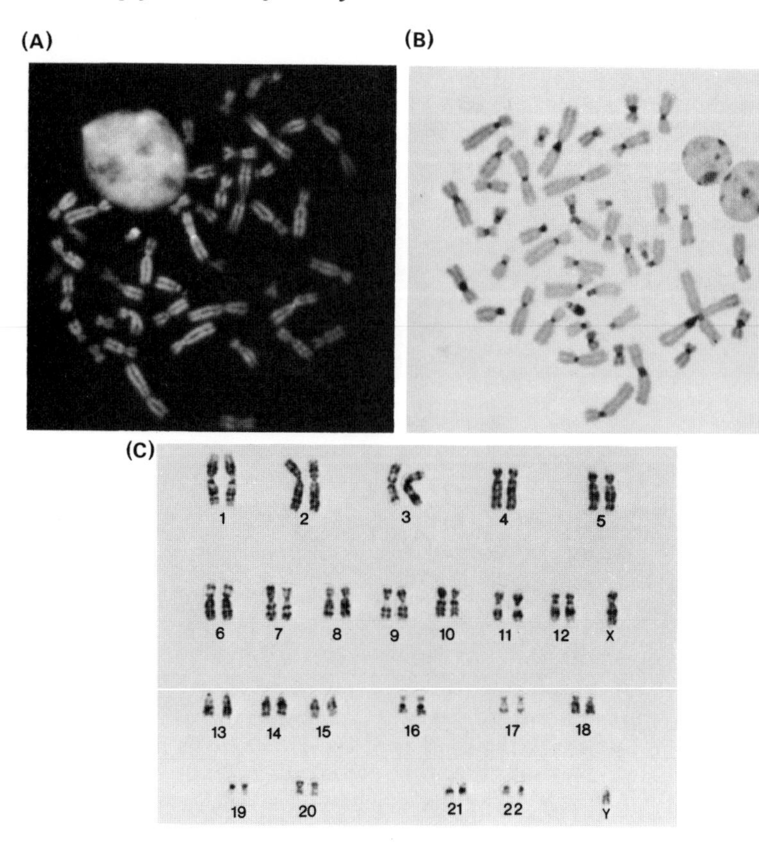

Fig. 1.5 Metaphase chromosomes of normal human males prepared from cultured blood lymphocytes: (A) Staining with quinacrine to show fluorescent Q-bands. (B) C-banding to demonstrate the location of constitutive heterochromatin. (C) Staining with acetic acid–saline–Giemsa to produce G-banding. [Courtesy of H. J. Evans]

matter in the vicinity of the centromeres. Densely stained bands of different size are interspersed with lightly stained interband segments. As we shall explain later the dark bands correspond with the sites of genes mapped by breeding tests. While the order of bands in these amplified (polytene) chromosomes is constant there is puffing at sites of transcription due to the RNA synthesized. Puffs develop in a regular sequence during larval development, partly at least in response to the increasing concentration of the hormone ecdysone. They mark directly the induction of transcription at specific loci (Ashburner, 1969).

Puffing in polytene nuclei is a reflection of the unravelling and spinning out of chromosome fibrils. A similar phenomenon is observable in the diplotene chromosomes of amphibian oocytes. These diplotene chromosomes are enormously long, ranging from 350 to 800 μm in

Triturus viridescens. As shown in Fig. 2.1 the varying but characteristic number of loops are spun out at specific sites (chromomeres) from the chromosomes along the axes of the bivalents. The pattern of loops which gives rise to the lampbrush appearance of the chromosomes is characteristic of the species. There are occasional mutants where the looping pattern is altered. These mutations are readily detectable in heterozygotes. More important, the fact that the structural difference with respect to loops is preserved in heterozygotes serves to emphasise that the looping pattern is a permanent structural feature of individual chromosomes. We shall consider lampbrush loops in more detail later in relation to both the molecular organization of the chromosomes and their genetic activity. In the meantime we note that they provide most detailed markers throughout the amphibian complement.

The differential coiling of segments in lampbrush and polytene nuclei finds a parallel in the chromomeres which give a beaded appearance to prophase chromosomes, especially at pachytene of meiosis. It is important to note, however, that while the chromomere pattern of prophase is specific and characteristic of a species it may vary substantially from one developmental stage to another. For example there is often a sharp reduction in the number of chromomeres at prophase of the second division of meiosis as compared with the first division, e.g. in *Agapanthus* (Lima-De-Faria, 1954). Apart from anything else the distinction serves to emphasize the changing character of the chromosome phenotype in relation to differentiation and development. It also warns against drawing conclusions from comparisons between chromomeric and banding patterns without taking account of variation due to differentiation.

Chromosome phenotype and genotype

Implicit in the discovery that the genes and other units of genetic information are located within the DNA of the chromosomes is the premise that the genotype of any individual is defined by the order of base sequences within the members of the chromosome complement. With rare exceptions (see Chapter 2), this order is preserved throughout the growth and development of each organism. As we have seen, however, the appearance and composition of chromosomes, leaving aside the composition of their DNA, is by no means constant during development. Neither is their behaviour. The chromosomes of salivary gland cells of *Drosophila* are very different from those of brain cell chromosomes. At mitosis the sequence of events and their consequences are very different from those at meiosis. Even in one tissue, such as the root meristem of seedlings in plants, the size and mass of chromosomes increase by as much as threefold from germination to three weeks (Bennett and Rees, 1969). Constancy, therefore, applies to chromosomal DNA only. In other respects the appearance and behaviour of chromosomes changes with changing cell environment. Chromosomes, in other words, have a well-defined phenotypic as well as genotypic aspect. From this we should expect that phenotypic changes in chromosome form and behaviour would be influenced by changes in both the external environment and by changes in genes which serve to control the activities of chromosomes. This, of course, is precisely what we find. There is ample evidence to show the dependence of all aspects of chromosome behaviour upon the genotype of the nucleus and, equally, upon the external environment of the organism. That the chromosome phenotype comes under the control of genes is clearly of significance from an evolutionary standpoint because it provides the means for adaptive changes affecting the transmission and distribution of genetic information (Chapter 3).

1.3 Prokaryote chromosomes

In the plant viruses and the smaller animal viruses the chromosomes comprise a single strand of RNA sometimes linear, sometimes circular (Table 1.1). In the larger animal viruses the chromosomes are of DNA, either single stranded or double stranded like the chromosomes of bacteria. As we might expect, the size of the nucleic acid molecule varies between the species but, without exception, the chromosomes consist of a naked nucleic acid thread lacking the histone and other proteins which comprise a substantial fraction of the chromosomes of eukaryotes.

One striking feature of prokaryotes is that chromosomes accept insertions of DNA segments incorporating genes derived from the same or from different species. The sex factor (*F*) in *E. coli* which promotes conjugation between bacteria is an example. This may exist as a particle in the cytoplasm which is independent of the chromosome or inserted to become part of the DNA helix of the normal chromosome. Such particles are called episomes. Their insertion may be facilitated by the 'naked' condition of the chromosomal DNA of prokaryotes. It may well be, on the other hand, that similar bodies are found in eukaryotes, but that they are more difficult to detect. Indeed Zhdanov (1975) provides good evidence to indicate the incorporation of the measles virus into the chromosomes of man. The measles virus is thought to interact with a latent oncornavirus, resulting in transcription of the measles virus RNA into double-stranded DNA which then integrates into the human genome.

2

Fine structure and organization

2.1 Architecture

Mononemy

Electron micrographs show that the chromosomes of eukaryotes are composed of fibres. These vary in diameter from one phase of the cell cycle to another and from one tissue to another. They consist mainly of protein and DNA. The structural organization of these fibres within the chromosome is interpretable as follows.

1 Each fibre contains one DNA helix in association with protein.
2 Variation in fibre diameter is dependent upon the density of coiling.
3 Each chromosome is made up of one fibre, and therefore one DNA helix, running continuously from one end of the chromosome to the other. Each chromosome in other words is *mononemic*, rather than *polynemic*; single-stranded rather than multi-stranded.

Compelling evidence for mononemy comes from measurements of lampbrush, diplotene chromosomes in the oocytes of the newt *Triturus viridescens* and from measurements of the DNA molecules extracted from *Drosophila* cells.

1 In *Triturus* each chromatid is represented by a single fibre (Fig. 2.1). When the protein is digested away by trypsin the diameter of the fibre is 2–3 nm. This would accommodate one, and only one, DNA helix (Miller, 1965).
2 Cells of wild-type *Drosophila* lysed with detergent and digested with pronase, yield chromosomal DNA molecules which are more or less intact, i.e. unfragmented. The mass of the largest DNA molecule in lysates corresponds closely with that of the DNA content of the largest chromosome of the complement. Furthermore, in a mutant stock where chromosome 3 is increased in length due to the attachment, by translocation, of a large piece of the *X*-chromosome, the largest DNA molecule increases in mass in direct proportion to the DNA increase in chromosome 3 (Table 2.1).

Packing and coiling

In the interphase nuclei of man the chromosome fibres have a dia-

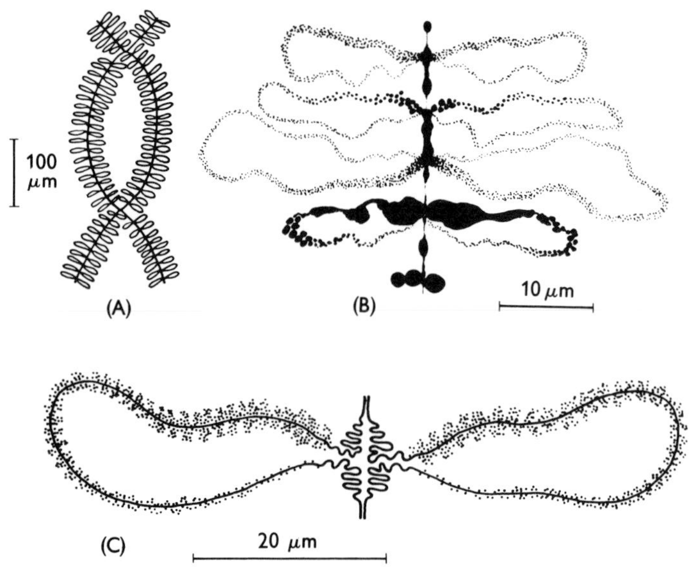

Fig. 2.1 Lampbrush chromosomes in *Triturus viridescens*. (A) A bivalent showing homologous chromosomes joined together by two chiasmata. (B) The central chromosome axis with paired lateral loops, showing differences in loop morphology and length. (C) The single stranded chromatids in the main chromosome axis are continuous with those of the lateral loops. [From Gall, J.G. (1956) *Brookhaven Symp. Biol.*, **8**, 17–32]

meter of approximately 23 nm and a dry mass of 6.1×10^{-16} g/μm. Thirty per cent or more of the mass consists of DNA. An extended DNA helix 1 μm in length weighs only 3.26×10^{-18} g. Clearly the DNA within the fibre must be tightly packed. The 'packing ratio' (extended double helix : packed double helix) has been estimated at 56 : 1 (DuPraw, 1970). The packing is probably achieved by compaction of the DNA within the fibre and the supercoiling of the fibre itself. Variation in the degree of compaction and of supercoiling of the fibre accounts for the variation in diameter. The indications are that coils and supercoils are maintained by histone proteins bound at specific sites to the DNA helix. The compactness of chromosomes at metaphase is achieved in two ways: (1) by

Table 2.1 Molecular weights and DNA contents of *Drosophila* chromosomes. [Data of Kavenoff, R., Klotz, L. C. and Zimm, B. H. (1973) *Cold Spring Harb. Symp. quant. Biol.*, **38**, 1–8]

Stock	M.W. of largest DNA molecule	DNA content of largest chromosome
Wild type	$41 \pm 3 \times 10^9$	43×10^9
Translocation	$58 \pm 6 \times 10^9$	59×10^9

contraction, roughly two-fold, in fibre length. This is clear from the increased DNA packing ratio of metaphase (100:1) in comparison with interphase fibres (56:1); (2) by folding and closer packing of the fibres themselves.

Recent work indicates that the compaction of the nucleoprotein (chromatin) fibre is, in the first instance, intermittent. Olins and Olins (1974) and Oudet, Gross-Bellard and Chambon (1975) have produced electron micrographs of fibres from interphase nuclei of the rat and the chicken which show clumps of chromatin separated by a fine thread. The clumps, called nucleosomes by Oudet *et al.* are 12.8 nm in diameter. The thread has a diameter of 1.5–2.5 nm. Each repeat in this string of beads contains about 200 DNA base pairs. The pattern, according to Kornberg and Thomas (1974) is determined by histone linkages with the DNA. Their evidence comes from mixing the tetramer of histone H3 and H4 along with the dimer of H2A–H2B (Lewin's nomenclature, 1975) with DNA *in vitro*. The X-ray diffraction pattern of the DNA/protein complex

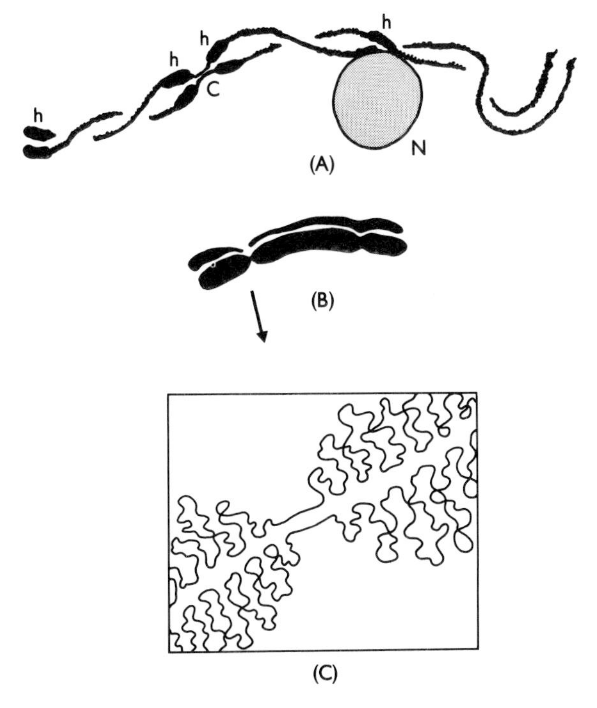

Fig. 2.2 Coiling and supercoiling at different stages of mitosis. (A) prophase showing localized densely coiled heterochromatin, h, at one end, on either side of the centromere, C, and in the region of the nucleolus organizer, N. (B) highly contracted chromosome at metaphase with chromatids unseparated in the vicinity of the centromere. (C) the fibrillar organization at the centromere, each chromatid comprising one highly convoluted and compacted fibre.

produced was identical with that of chromatin. The fifth histone, H1, and the non-histone proteins are not involved. This is not to say, however, that they play no part in compaction at a higher level, by coiling and supercoiling. The coiling and supercoiling properties of chromosomes at different stages of mitosis are illustrated in Fig. 2.2.

DNA and genes

In *E. coli* the chromosomal DNA is made up of about 4 000 000 base pairs. Approximately 1000 base pairs provide the information which specifies the protein or polypeptide chain synthesized under the control of each gene. On this basis the *E. coli* chromosome carries 4000 genes, assuming each gene to be represented by a single copy. This is almost certainly an overestimate because recent evidence suggests that the RNA messenger transcribed at each locus contains base pairs supplementary to those providing information about the amino acid sequence of the polypeptide product. Even so 4000 is an acceptable estimate and most would agree also with the generalization that the *E. coli* chromosome is made up of genes, side by side, each in single or very few copies. In eukaryotes the DNA content of chromosome complements is far greater than in *E. coli* and other prokaryotes (Table 2.2). The increase in DNA content is not surprising on grounds of increasing complexity and consequent demand for increase in the amount of information, in other words for a greater variety of genes and gene products. What is surprising, however, is the magnitude of the DNA variation between prokaryotes and eukaryotes and even within eukaryotes. For example the diploid nuclei of man, at early interphase, contain 6.4×10^{-12} g of DNA (equivalent to 5800 million base pairs as compared to 4 million in *E. coli*). If each allele is represented by a single copy the diploid complement in man is made up of almost 6 million genes, the haploid complement of half that number. Even allowing for the complexity and elaboration of eukaryote cells relative to prokaryotes the estimate is unreasonably high and human cells are by no means amongst those with the highest DNA contents (Table 2.2). The conclusion is reinforced by the wide variation in DNA amount between closely related species. *Vicia faba* has seven times as much DNA as its close relative *V. sativa*. Both are diploids and it is clearly unreasonable to assume the chromosomes of the former incorporate seven times as many kinds of genes as the latter. To account for the large amounts of DNA in eukaryote nuclei and for the magnitude of the DNA variation there are a number of possibilities:

1 A number of genes, i.e. of informative and transcribable DNA sequences, are highly repeated throughout the complement.
2 A number of sequences, either in single copies or in many copies are uninformative and not transcribable.
3 Some of the variation in DNA content reflects differences in the number and variety of genes in different species but, as we have

Table 2.2 Amounts of DNA in eukaryotes and prokaryotes (pg)

Eukaryotes		
ANIMALS	DNA per 2C nucleus	2*n*
Amphiuma	168.0	24
Protopterus (Lungfish)	100.0	38
Salamandra salamandra	85.3	24
Triturus viridescens	72.0	22
Bombina bombina	28.2	24
Rana esculenta	16.8	26
Rana temporaria	10.9	26
Bos taurus (ox)	6.4	60
Man	6.4	46
Sheep	5.7	54
Mouse	5.0	40
Drosophila	0.2	8
FLOWERING PLANTS		
Fritillaria davisii	196.7	24
Lilium longiflorum	72.2	24
Tulipa gesneriana	51.5	24
Allium cepa	33.5	16
Vicia faba	28.0	12
Lathyrus latifolius	21.8	14
Ranunculus ficaria	19.2	16
Secale cereale	18.9	14
Zea mays	11.0	20
Crepis capillaris	4.2	6
Vicia sativa	4.0	12
Antirrhinum majus	3.6	16
Linum usitatissimum	1.4	30
FUNGI	DNA per haploid nucleus	*n*
Dictyostelium discoideum	0.384	7
Ustilago maydis	0.208	2
Aspergillus nidulans	0.048	8
Saccharomyces cerevisiae	0.026	15
Prokaryotes		
BACTERIA	DNA per cell	
Salmonella typhimurium	0.0143	
Escherichia coli	0.0040	
Diplococcus pneumoniae	0.0022	
Mycoplasma hominis	0.0009	
VIRUSES		
T2	0.000 220	
λ	0.000 055	
P4	0.000 016	
ϕX174	0.000 005	

indicated, this is unlikely to account for more than a small fraction of the variation observed.

2.2 Repeated sequences

Callan and Lloyd (1960), on the basis of their observations of lampbrush chromosomes in *Triturus* oocytes, were among the first to propose that genes within chromosomes were represented by numerous copies. Their proposal is based on the premise that each chromomere in the lampbrush chromosome corresponds with the location of a gene. At each chromomere a loop is spun out and it is on this loop that the RNA messenger is transcribed and protein synthesized (see Chapter 1). The evidence for the latter comes from Gall and Callan (1962) who showed that tritiated uracil (in RNA) and tritiated phenylalanine (in protein) were incorporated by the chromomere loops and only by the loops. The average loop, however, is at least 150 000 nucleotides in length, many times longer than the 1000 nucleotides required for the synthesis of the average polypeptide. The conclusion was that each gene was represented by many copies in tandem.

A number of methods are now available to confirm the high degree of repetition within chromosomal DNA.

Cot curves

DNA extracted in its normal, 'native', double-stranded form, sheared into short fragments, e.g. by sonic disintegration, rendered single-stranded (denatured) by heating or by strong alkali can be renatured, i.e. restored to the double-stranded condition, under suitable conditions of temperature, pH and cation concentration. The rate at which the DNA renatures is dependent upon the genome size, i.e. the variety of differing base sequences and the extent to which any of the sequences are repeated. Only complementary single strands will pair and anneal and clearly the greater the repetition of sequences the greater the likelihood of collisions between complementary strands. Another factor which influences the rate of renaturing is obviously the concentration of the DNA: the greater the concentration the more frequent the collisions. For this reason the rate at which the DNA renatures is plotted as a percentage against concentration × time, expressed in moles of nucleotides × seconds per litre (*Cot*). The percentage DNA renatured at a particular *Cot* value may be established in a number of ways (Britten and Kohne, 1968). One is to pass the DNA in solution through a hydroxyapatite column which traps that fraction of the DNA which has renatured, i.e. the double-stranded DNA. Another is to measure the optical density of the DNA by spectrophotometry. Single-stranded DNA absorbs more ultraviolet than double-stranded DNA. With renaturing therefore the hyperchromicity decreases in proportion to the amount of double-stranded DNA. In

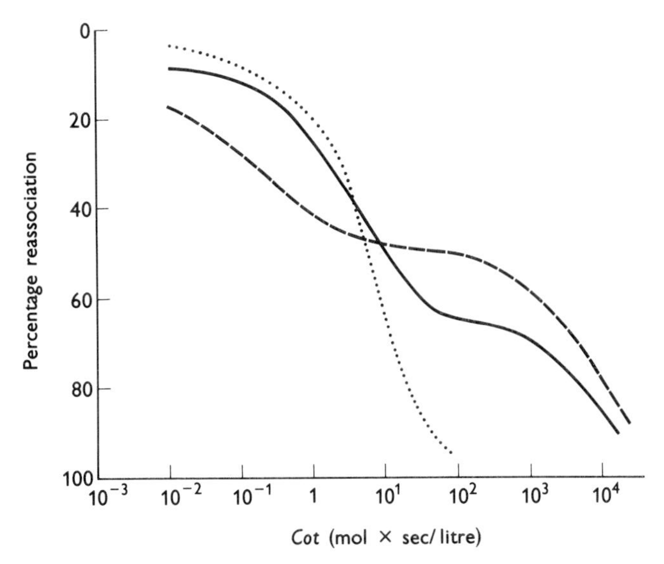

Fig. 2.3 *Cot* reassociation curves for the DNA of two eukaryotes, *Lathyrus sativus* (———) and the calf (- - - -), and of the prokaryote *E. coli* (. . . .).

Fig. 2.3 are the *Cot* curves for *E. coli*, the calf and the plant species *Lathyrus sativus*.

The distribution and dimensions of repeated sequences

Despite the fact that the variety of base sequences within the complement of eukaryotes is very much greater than in *E. coli* it will be observed in Fig. 2.2 that the percentage of the DNA of the eukaryotes (calf and *Lathyrus sativus*) renatured at low *Cot* (10^{-3} to 1) is in fact greater than for *E. coli*. This rapid renaturing at low *Cot* shows that some of the base sequences in the eukaryotes are highly repetitive. Between 1 and 30% of the DNA of the eukaryotes renatures at very low *Cot* values, up to 10^{-2}. This represents, on an arbitrary basis, the 'fast' reassociation component, the highly repetitive fraction of the DNA, sequences repeated 10 000 times or more (Southern, 1974). Beyond *Cot* 10^{-2} the moderately repeated sequences and the single copy or unique sequences are renatured. The units of repetition range in length from as few as six base pairs in the guinea pig (E. M. Southern, 1970) to several hundred.

Satellite DNA

When native DNA is centrifuged in a density gradient such as caesium chloride the buoyant density at equilibrium is dependent upon its G + C content. The higher the G + C content the higher the buoyant density. The DNA of eukaryotes has, on average, a G + C content of 30–50%. The G + C content, however, varies from one section of the DNA mole-

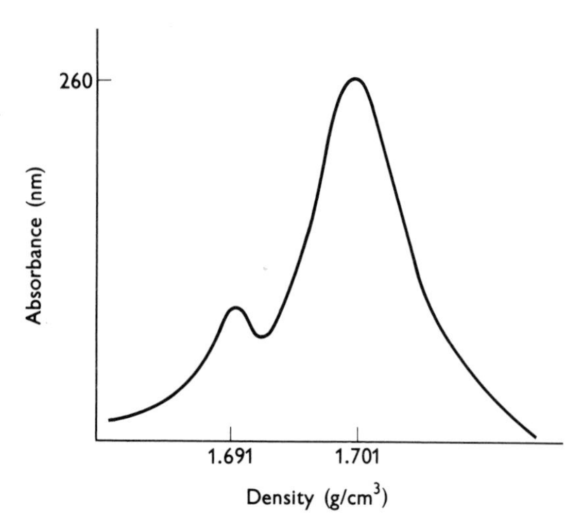

Fig. 2.4 The plot of UV light absorbance against density for mouse DNA centrifuged in a caesium chloride density gradient. The satellite DNA peak (density 1.691 g/cm^3) is on the light side of the main band DNA (density 1.701 g/cm^3). [From Evans H. J. and Sumner A. T. (1973) *Chromosomes Today*, **4**, 15–33]

cule to another by 10% or more. For this reason the sedimentation coefficient of the DNA from one species covers a range of buoyant densities, giving a broad band when plotted in a buoyant density curve. Exceptional DNA segments may have an unusually high or low G+C content. When plotted, these fractions appear as heavy or light satellites respectively at the tails of the 'main-band' DNA. Heavy satellites are found in the guinea pig and in human DNA. Light satellites, like the one shown in Fig. 2.4 for mouse DNA, are less common. One reported in the crab has a G+C content of only 5%; others have been identified in *Drosophila* species.

Highly repetitive segments with distinctively high or low G+C contents, relative to the main band DNA, would clearly be detectable as heavy and light satellites respectively. It is worth emphasizing, however, that a highly repeated sequence is not necessarily detectable as a satellite. For example, a highly repeated sequence with the same base ratio as the bulk of the DNA would have the same buoyant density and 'settle' in the same density band.

The mouse satellite represented in Fig. 2.4 makes up 10% of the total DNA. Its location within the mouse complement was established with considerable precision by annealing tritium-labelled single-stranded DNA or else radioactive RNA copied in vitro from mouse satellite DNA (complementary or cRNA) with chromosomal DNA in situ (K. W. Jones, 1970; Pardue and Gall, 1970). For this purpose chromosomal

DNA is rendered single stranded either by heating or by treatment with alkali before being brought into contact with the labelled DNA of the satellite or else labelled RNA complementary to it. Autoradiographs show that the satellite DNA and the cRNA bind to the heterochromatic segments around the centromeres and at the nucleolus organizers. Similar methods have established that the highly repetitive satellite DNA in *Drosophila* is also located mainly in the centromeric heterochromatin (Jones and Robertson, 1970).

Thomas' circles

An ingenious method developed and applied by Thomas and his colleagues shows that repetitive DNA segments are not necessarily contiguous within chromosomes. Fragments of DNA in double-stranded form are exposed to an exonuclease, e.g. exonuclease III, which digests single strands of the double helix in the 3′ to 5′ direction. When sufficient of the undigested single strands, at least 200 nucleotides in length, are exposed at each end, the ends with complementary base sequences will pair under conditions permitting renaturing to form stable rings (Fig. 2.5) readily distinguishable under the electron microscope (Thomas, 1970). Complementarity is, of course, evidence for repetition and the proportion of fragments forming rings is an index of the degree of repetition within the chromosomal DNA. Another factor which influences the capacity to form rings is the length of the fragment. In *Drosophila* the optimum length is 1.2 μm. With longer fragments ring formation is reduced, from which it is inferred that repetitive sequences

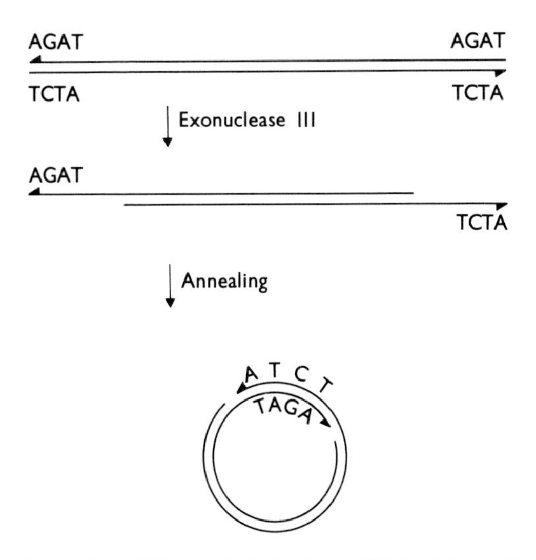

Fig. 2.5 The formation of Thomas' circles from DNA with tandemly repeating base sequences.

are in clusters, *g* regions, interspersed with unique sequence DNA. The average length of a *g* region is estimated to be 5 μm, i.e. about 22 500 nucleotides.

Twenty two per cent of the DNA in *Drosophila* is highly repetitive and located in the heterochromatic regions around the centromeres. For this reason ring formation is mainly restricted to the DNA in heterochromatin in this case (Hutton and Thomas, 1975; Peacock *et al.*, 1973; Schachat and Hogness, 1973). This is not to say there is no element of repetition elsewhere, interspersed within the euchromatic fraction of the complement.

Davidson *et al.* (1973) provide further detailed information on the distribution of repetitive sequences in animal DNAs. They incubated long strands of denatured DNA, labelled with tritium, with an excess of short, unlabelled single-stranded fragments at low *Cot*, i.e. in low concentration for a short period of time, such that only highly repeated sequences reassociated. The mixture was then passed over hydroxyapatite which binds to the duplexes made up of long labelled strands paired with short unlabelled fragments carrying complementary repeated sequences. Most of the bound, labelled strands included non-repeated as well as the repeated sequences with which the short fragments are paired. They concluded from this that repeated sequences throughout the genome are interspersed with unique sequences. In *Xenopus*, by using labelled strands of different length, they estimate that: (1) the majority of repetitive sequences are short, averaging 300 nucleotides; (2) each of the sequences repeated is separated by a unique DNA sequence 700–1000 nucleotides in length, some by unique sequence segments up to 4000 nucleotides; and (3) a small fraction of the total DNA, about 7 % is made up of repetitive sequences many thousands of nucleotides in length. Davidson *et al.* (1973) and Thomas *et al.* (1973) also, place particular emphasis on the alternating pattern of repeated and of unique sequences in the DNA of *Xenopus* and other animals. What this signifies in functional terms is not known.

The composition of repeated sequences

E. M. Southern (1970) and Gall, Cohen and Atherton (1973) have established in detail the dimensions and base composition of satellite DNAs in the guinea pig, and in *Drosophila*. In the guinea pig satellite the sequence is $\substack{\text{CCCTAA}\\ \text{GGGATT}}$ repeated, with some variation due to mutations, millions of times. Satellites I, II and III in *Drosophila virilis* comprise thousands of repeats of the sequences $\substack{\text{ACAACT}\\ \text{TGTTGA}}$, $\substack{\text{ATAAACT}\\ \text{TATTTGA}}$ and $\substack{\text{ACAAATT}\\ \text{TGTTTAA}}$ respectively. None of these sequences makes any sense from the standpoint of coding for protein. This is not to say, however, that repetition involves only nonsense sequences of DNA bases. One example of a transcribable base sequence repetition, to which we have already referred, is the sequence coding for ribosomal RNA. It will be recalled that in *Drosophila melanogaster* the wild type fly carries at the *bobbed* locus about 130 copies of the gene sequences which code for the 18*S* and 28*S* components of the ribosomal RNA (Ritossa and Spiegelman, 1965).

It has also been suggested that the bands in the *Drosophila* chromosomes, which correspond to the sites of single genes (Judd and Young, 1973), may represent many copies of each gene. In DNA terms each band on a chromosome includes some 25 000–50 000 nucleotide pairs, 'sufficient' for 20 or more copies of a gene. The evidence of Peacock *et al.* (1973) is against this view on the grounds that most of the repetitive material, they claim, is concentrated in the centromeric heterochromatin, rather than in the bands. One may speculate that the DNA extra to the transcribed sequences within the bands may fulfil regulatory or other functions. The truth is that one does not know.

Chromosomes and information

Until fairly recently the chromosomal DNA was considered, in the main at least, to be made up of gene sequences in tandem much like the beads on a string. For prokaryotes the concept is still largely valid. The picture which now emerges for eukaryotes is, as we have indicated, much more complex. Much of the DNA comprises a variety of untranscribable sequences, some of which are repeated many thousands of times. Many transcribable gene sequences are repeated, others are represented by single unique DNA sequences. Some of the unique sequence DNA is untranscribed, like the 'spacer' sequence between genes coding for ribosomal RNA in *Xenopus laevis* (Loening, Jones and Birnstiel, 1969). It needs to be emphasized that from a genetical as well as structural standpoint, the chromosome in heredity represents very much more than a row of genes coding for polypeptides.

2.3 Replication and division

Mitosis in eukaryotes

In a tissue of multiplying cells a proportion of nuclei are in interphase, the remainder in division. The ratio of one to the other, as we shall see, is of use in timing the sequence of events during the mitotic cycle. From the genetic standpoint the characteristic and significant consequence of mitosis (Fig. 2.6) is that the products of division are identical, embodying precisely the same genetic material as the parent nucleus. The justification for this statement comes from many sources. On cytological grounds we observe that the chromosomes in the two nuclei produced by division have the same appearance and the same DNA content as the parent nucleus. On genetic grounds we observe that the offspring of asexual reproduction, whether by budding or clonal propagation, have similar genotypes. A further compelling proof is that when the nucleus from a specialized epithelial cell of the intestine of a *Xenopus* tadpole is transplanted into an unfertilized egg whose nucleus has been removed or inactivated, the egg can develop to produce a normal adult. We con-

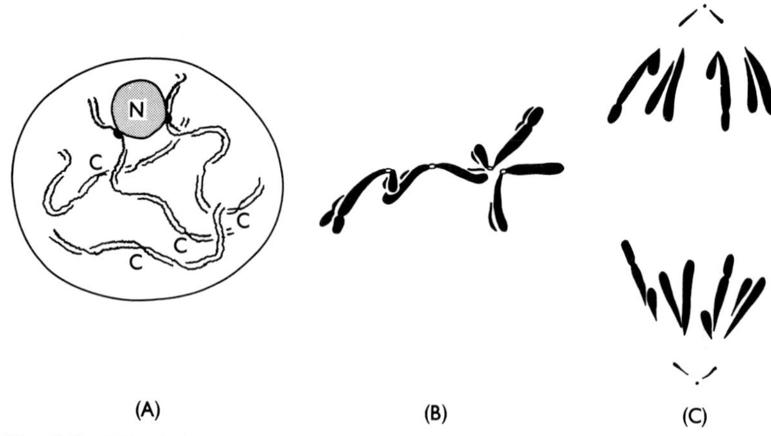

Fig. 2.6 Mitosis in an organism with two pairs of chromosomes. (A) Prophase: each chromosome comprising two chromatids. C = centromere. N = nucleolus, attached to the nucleolus organizer regions. (B) Metaphase with centromeres aligned along the equator of the spindle. (C) Anaphase showing equal separation of daughter chromosomes to opposite poles of the cell.

clude that the information embodied within the chromosomes is preserved unchanged during the many cycles of mitotic division leading up to the formation of the larval intestine (Gurdon, 1973). There are, as we shall consider below, certain exceptions to this general rule.

The mitotic cycle

Nuclei. Replication of chromosome material, including the synthesis of DNA, takes place during interphase. The onset and duration of the synthesis and of other phases of the cycle are readily established by autoradiography. The method for root meristems for example, is as follows.

1. Roots are immersed briefly in a solution containing tritiated thymidine and, after washing, returned to water or culture solution. The labelled thymidine is incorporated by the DNA synthesized in interphase nuclei.
2. Root tips are fixed at intervals of one hour over a period of 24 hours or more, then stained and squashed.
3. Coverslips are removed, the squashed cells on the slide covered with a photographic emulsion and the slides kept in darkness. After some ten days or so the emulsion is developed and fixed, after which the location of labelled DNA among nuclei and among chromosomes is established by the distribution of silver grains in the autoradiograph.

Timing the cycle and its components. The proportion of nuclei labelled at a well-defined stage, e.g. prophase, is plotted against time (Fig. 2.7). From this graph the following information is established.

The total cycle. The interval between the two peaks represents the average time for the complete mitotic cycle. The first peak in the graph represents the first batch of labelled nuclei to reach prophase, the second peak the nuclei entering prophase for the second time after labelling. The most efficient estimate of the interval between peaks is obtained from the difference between points half way up the ascending curve of the peaks (A and A' in Fig. 2.7).

The synthesis (S) *phase.* This is estimated from the interval, A–S, the midpoints on the ascending and descending curves. Another way of estimating S is from the proportions of nuclei labelled (say one hour after labelling) $\times$ the duration of the total cycle. The basis for this estimate is that the period spent at a particular phase of the total cycle is directly reflected by the proportion of nuclei at that phase. Thus, we expect to observe more nuclei at a particular phase if that phase is of long duration, and vice versa. The estimate holds good if the cells are unsynchronized, proceeding through mitosis independently of one another in random order. This holds true for most somatic tissues in adult organisms, including root meristems.

The post-synthesis phase ($G2$). This is given by the interval from the time of immersion in tritiated thymidine to A, the average time at which cells enter prophase.

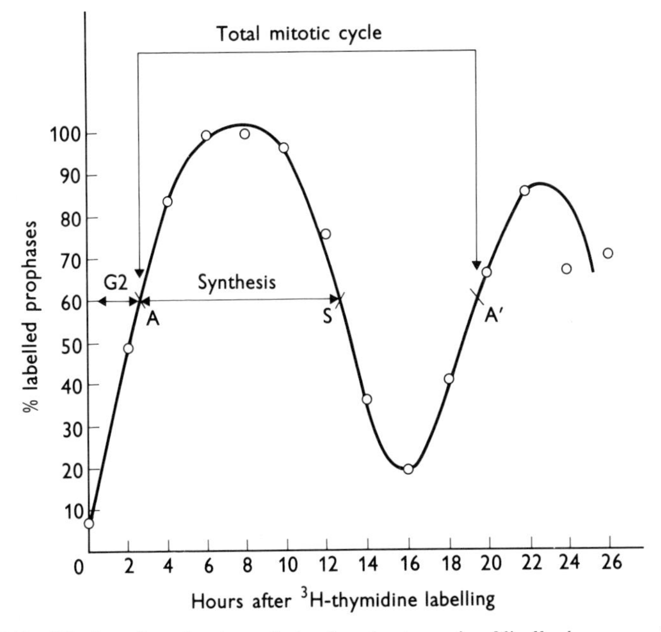

Fig. 2.7 Timing the mitotic cycle in the plant species *Nigella damascena*. [Data from Evans, G. M., Rees, H., Snell, C. L. and Sun, S. (1972) *Chromosomes Today*, **3**, 24–31, with permission]

The division phase (D). The proportion of cells in division, prophase to telophase, reflects directly the proportion of the total cycle time spent in division. The duration of component phases of the division may be estimated in the same way.

The pre-synthesis phase ($G1$). $G1$ is estimated by the difference, total cycle time minus the duration in division, $G2$ and S.

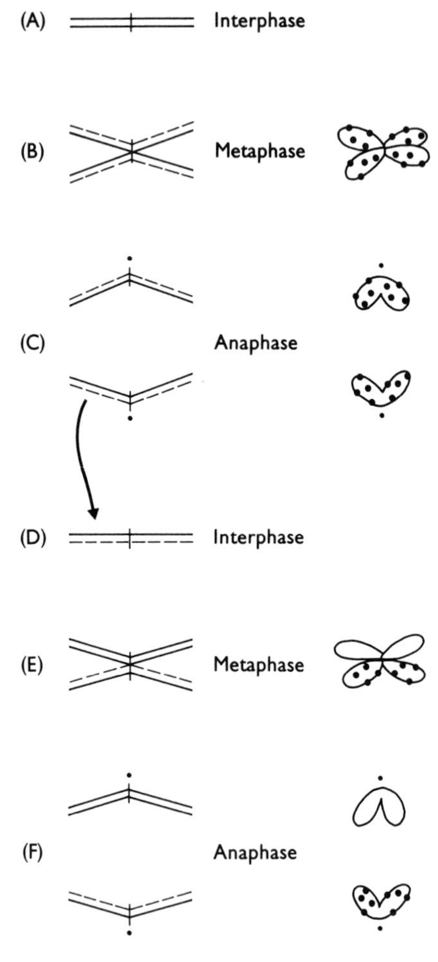

Fig. 2.8 Taylor's experiment. Following replication of the double helix (at A) in the presence of tritiated thymidine one of the DNA strands in each of the chromatids at metaphase (B) and anaphase (C) is labelled. As a result both chromatids produce silver grains in autoradiographs (right). After replication of the daughter chromosome (D) in the absence of tritiated thymidine only one of the chromatids contains a labelled DNA strand at the following metaphase (E) and anaphase (F).

The procedure for timing the mitotic cycle and its component phases is illustrated with respect to the root tip meristems of *Nigella damascena* in Fig. 2.7. The timing of the mitotic cycle is influenced by environmental factors, especially temperature, so that for purposes of comparison it is clearly important to standardize conditions of culture and experimentation. We shall be considering in more detail later some of the genetic factors which influence the duration of the mitotic cycle and its components.

Chromosome duplication

An experiment by Taylor, Woods and Hughes (1957) marked a very considerable advance in our understanding of the replication of chromosomes. After labelling the DNA during interphase with tritiated thymidine, both the chromatids of each chromosome at the subsequent metaphase carry the label. At the second metaphase following label one of the two chromatids is labelled (Fig. 2.8). The results are precisely those expected on the view that each unreplicated chromosome (chromatid) is mononemic, i.e. containing one DNA double helix which, of course, replicates semi-conservatively.

In prokaryotes such as *E. coli* labelling with tritiated thymidine also showed a semi-conservative replication of the chromosomal DNA. These experiments also demonstrated that replication proceeds bi-directionally from a single fixed point on the chromosome. In eukaryotes replication is also bidirectional but proceeding from a number of initiation points within each chromosome. In somatic cells of *Xenopus* the initiation points are spaced at intervals of 20–125 μm, in *Triturus* 125 μm or more (Callan, 1972). The duration of the DNA synthesis phase varies however between tissues. The reason is a change in the spacing between initiation points. In *Triturus* for example the *S* phase in spermatocytes, relative to differentiated somatic cells, is protracted and the number of initiation points reduced by wider spacing along the DNA helix. The *rate* at which replication extends from the initiation points remains constant. The same correlation between the duration of *S* in different tissues and of the intervals between initiation points applies in *Drosophila* (Blumenthal, Kriegstein and Hogness, 1973). Change in the rate of DNA synthesis between initiation points could also affect the duration of *S* between cells of different tissues but we know of no evidence of control by this means.

DNA synthesis and replication in euchromatin and heterochromatin

The synthesis of DNA is not synchronous throughout the chromosome complement. The most marked difference in the timing of DNA synthesis is that between euchromatin and heterochromatin. This was first established by Lima-De-Faria (1959). After treatment with tritiated thymidine the heterochromatin was heavily labelled late in the synthesis (*S*) phase when the euchromatin carried little or no label, showing that the synthesis continued in heterochromatin after cessation of DNA synthesis in the euchromatin.

The distinction between heterochromatin and euchromatin may extend to the replication and transmission of chromosome material during cell division. Normally during mitosis the products of division are genetically identical, with each other and with the parent nucleus. There are startling exceptions. In *Sciara coprophila* whole chromosomes, the L-chromosomes, are eliminated from somatic cells at about the fifth cleavage division. Three or four divisions later one of the three X-chromosomes in the female, and two of the three in the male are also eliminated from all but germ line cells. Both the L- and X-chromosomes are largely heterochromatic. Another class of 'dispensable' chromosomes, the E-chromosomes of dipterans, is shed from somatic nuclei. Like the Ls and Xs they are also heterochromatic (White, 1973; Lewis and John, 1963).

Heterochromatic segments, making up some 60% of the chromosomes, are shed from somatic nuclei during cleavage in *Parascaris* (see White, 1973). In the polytene nuclei of the salivary glands of *Drosophila* the centromeric (α-)heterochromatin, which makes up about 30% of most somatic nuclei, is relatively much reduced. While the euchromatin replicates normally during the differentiation of the salivary glands the α-heterochromatin fails to reproduce itself. In the oocytes of amphibia, e.g. *Xenopus laevis* (Gall, 1968), a specific chromosome segment, the nucleolus organizer coding for ribosomal RNA, is amplified relative to other segments, the amplified products in this case being detached from the chromosomes and migrating to the nuclear envelope where they eventually lead to the formation of multiple nucleoli.

The products of mitosis

These exceptional cases of non-conformity in the replication of chromosome material within nuclei are significant for a number of reasons. First, they bear testimony to the dispensability of some of the genetic material, albeit in certain tissues and organs. Second, they establish the capacity for modification of the mitotic process to generate gain or loss of specific chromosome segments. In the examples we have given the loss or gain of segments is certainly specific in distribution within the nucleus and equally specific in timing during growth and development; in short the events are well ordered and controlled. Third, the possibility is that less obvious loss or gain, involving a very small fraction of the chromosome material, may well be a not uncommon phenomenon during development. Indeed Pearson, Timmis and Ingle (1974) have confirmed a widespread and substantial variation in the satellite fraction of the DNA located in the heterochromatin of Melon chromosomes in different tissues. Exactly what this signifies in functional terms awaits a better understanding of the role and properties of satellite and other dispensable DNA fractions. While it is right therefore to emphasise the uniformity of the products of mitosis in general it is equally important to emphasise the capacity for qualitative modifications of the chromosome complement by amplification or deletion especially within differentiated, as distinct from meristematic tissues.

Variation in the quantity of genetic material, by polyploidy or endopolyploidy is, of course, common in differentiated tissues of both plants and animals which provide, in themselves, further examples of adaptive modifications in the mitotic process. Genetic quality and constancy of the products are, therefore, a characteristic but not invariable consequence of mitosis.

3

Meiosis and recombination

3.1 The meiotic cycle

At mitosis in eukaryotes the synthesis of DNA and of the chromosome constituents is succeeded by the separation of chromatids at anaphase during the cell division which follows. The two daughter cells produced contain, as a result, identical sets of chromosomes which, in turn, are identical with those of the parent cell. At meiosis, in contrast, one round of DNA replication and chromosome duplication is followed by two successive nuclear divisions. At anaphase of the first of these divisions the homologous chromosomes are separated to opposite poles. At anaphase of the second division the chromatids are separated. This results in four cells, each with a reduced haploid complement which, in heterozygotes, differ from one another due to segregation and recombination. Fusion of the haploid male and female gametes at fertilization is, of course, the mechanism which restores the complement to the diploid state in the zygote.

DNA synthesis takes much longer at meiosis than at mitosis in most dividing tissues. In *Lilium* pollen mother cells DNA synthesis at meiosis takes about three times as long as in root meristem cells undergoing mitosis. DNA synthesis at meiosis in the testis of mouse and of *Triturus* is also much protracted relative to the duration of DNA synthesis in tissues undergoing mitosis. As we have mentioned earlier the longer duration of DNA synthesis at meiosis is associated with a reduction in the number of initiation points of synthesis within the DNA molecules.

Apart from a difference in the duration of DNA synthesis there is, at meiosis, a fraction of the DNA (about 0.3% in *Lilium*) that remains unreplicated at interphase. It is replicated during early prophase, at zygotene. This fraction is specific and distinguishable from the bulk of the nuclear DNA in having a higher G + C content: 50% as compared with 40% in the total DNA.

Following the *S* phase, *G2* at meiosis is very short or even nonexistent. The meiotic cycle embracing both divisions, also takes much longer than the mitotic cycle (Table 3.1). In man, for example, the average duration of the mitotic cycle in cultured Hela cells is about 24 hours. Meiosis in the spermatocytes takes about 24 days. In the female, meiotic divisions commence before birth. They are arrested at early diplotene of the first division until the eggs are shed such that the completion of meiosis, to give egg cells and polar bodies, awaits puberty at the earliest and may take 30 years or more. In flowering plants, as well, the meiotic cycles are generally of much greater duration than the mitotic cycles, from 1 to 11 days. As Bennett (1971) has shown the duration is closely correlated with the amount of nuclear DNA, as is the case for mitotic cycles.

While the facts relating to the differences in duration of mitotic and meiotic cycles are, in themselves, indisputable it is fair to say that the significance of such differences remains conjectural. There is a great deal of ignorance concerning the mechanisms and processes that make up the meiotic cycle.

Table 3.1 2C DNA amounts and the mitotic and meiotic cycle times in plant species. [Cycle time data from Bennet, M. D. (1971) *Proc. R. Soc. B.*, **178**, 277–99]

Species	$2n$	DNA (pg)	Cycle times (hours)	
			Mitosis	Meiosis
Haplopappus gracilis	4	3.4	11.9	24.0
Secale cereale	14	18.9	12.75	51.2
Allium cepa	16	33.5	17.40	96.0
Tradescantia paludosa	12	36.6	20.0	126.0
Lilium longiflorum	24	72.2	24.0	192.0
Trillium erectum	10	74.4	29.0	274.0

3.2 The meiotic divisions

The general sequence of events during division is similar throughout most diploid eukaryote species even though, as we have mentioned, the timing and duration may vary from one species to another. The following is a brief summary of the events at division. They are also presented in Fig. 3.1. We shall deal in greater detail with certain of these events in a later section.

The first division

Division commences directly or very soon after the completion of DNA synthesis. The precocity of the onset of prophase, as compared with mitosis, may explain why the chromosomes although duplicated in respect of their DNA content at the *leptotene* stage, still appear as single threads, i.e. without evidence of replication into chromatids. At *zygotene* the intimate pairing of homologous chromosomes begins. Darlington (1937) attributes the capacity for pairing to the single-stranded organization of the chromosomes at early prophase of meiosis, which in turn is attributable to the precocity to which we have referred. Pairing can be

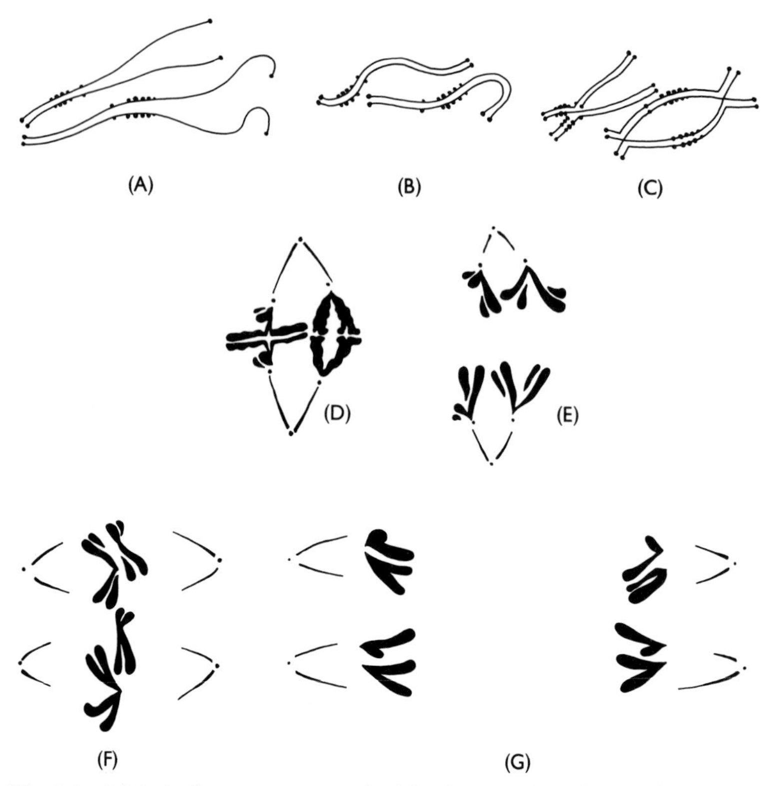

Fig. 3.1 Meiosis. Commencement of pairing between homologous chromosomes at zytogene (A); completed at pachytene (B). Chiasmata in diplotene bivalents (C) and at first metaphase (D). Separation of half bivalents at first anaphase (E), their orientation at second metaphase (F), and separation at second anaphase (G).

seen to commence at well-defined sites in many species, e.g. at the centromeres in *Fritillaria meleagris*, at the ends of the chromosomes in rye. It is probable that a similar localized initiation applies to all species. Pairing is specific in respect not only of homologous chromosomes but to homologous segments within the chromosomes. It is completed at *pachytene*. During pachytene the chiasmata are formed, marking sites of crossovers. The homologous chromosomes are held together by the chiasmata so that the complement is deployed as a set of bivalents corresponding to the haploid number. A striking feature of pachytene chromosomes is the longitudinal differentiation into a series of chromomeres. They represent specific and diagnostic sites of aggregation of material due to intensive localized coiling. At *diplotene* the attraction between homologous chromosomes lapses, indeed there is evidence of a repulsion between them. They are held together, however, by chiasmata. Both chromosomes of each bivalent are clearly resolved as double-stranded in organization, i.e. comprising separate chromatids. The shape of each bivalent at diplotene and in subsequent stages, including metaphase, is determined by the number and distribution of its chiasmata. Progressive coiling and contraction, which is continuous throughout prophase, leads to diakinesis which marks the disappearance of the nuclear membrane and the detachment and dispersion of the nucleoli. The spindle forms and at *first metaphase* (MI) the bivalents become orientated on the equator with the centromeres of homologous chromosomes directed to opposite poles. At first anaphase (A1) the half bivalents are separated and there follows some uncoiling of chromosomes at *telophase* which continues in the interphase nuclei. At the interphase there is no DNA synthesis or chromosome replication. The content of each of the pair of nuclei remains as half bivalents, equivalent to the haploid number.

The second division

The *second prophase* proceeds, much as in mitosis, with increasing contraction to the *second metaphase* (MII), with the exception, in most cases, that the chromatids lie well apart and not in close proximity as at mitosis. At *second anaphase* the centromeres of each half bivalent move to the poles so that telophase and, subsequently, interphase nuclei contain a haploid set of single-stranded chromosomes.

There are, of course, variations on this general theme (see John and Lewis, 1965). In plants a common phenomenon is the omission of cell wall formation after the first division, e.g. in *Paeonia*.

Univalents

Where, for some reason, there is a failure of chiasma formation the unpaired univalents fail to move on to the spindle equator or do so much later than the bivalents. In the former case they frequently form micronuclei and are 'lost' at the first division. In the latter they may divide at first anaphase to be included as single-stranded, chromatids in telophase nuclei. At second anaphase, however, they are incapable of movement from the equator and are often 'lost' as micronuclei at this second division. The chiasmata are, therefore, essential not only in the context of crossing over and recombination but also for the regular distribution of chromosomes to the gametic nuclei (Fig. 3.2).

An alternative, either at the first or second anaphase, is that the univalent undergoes 'misdivision', i.e. splits transversely within the centromere. The result will be two chromosomes with terminal centromeres, telocentrics. In many organisms telocentrics are unstable in the sense that they are converted to iso-chromosomes following fusion of the centromeres from identical arms (Fig. 3.3).

3.3 Chromosome pairing

While observations under the light microscope confirm the fact and the

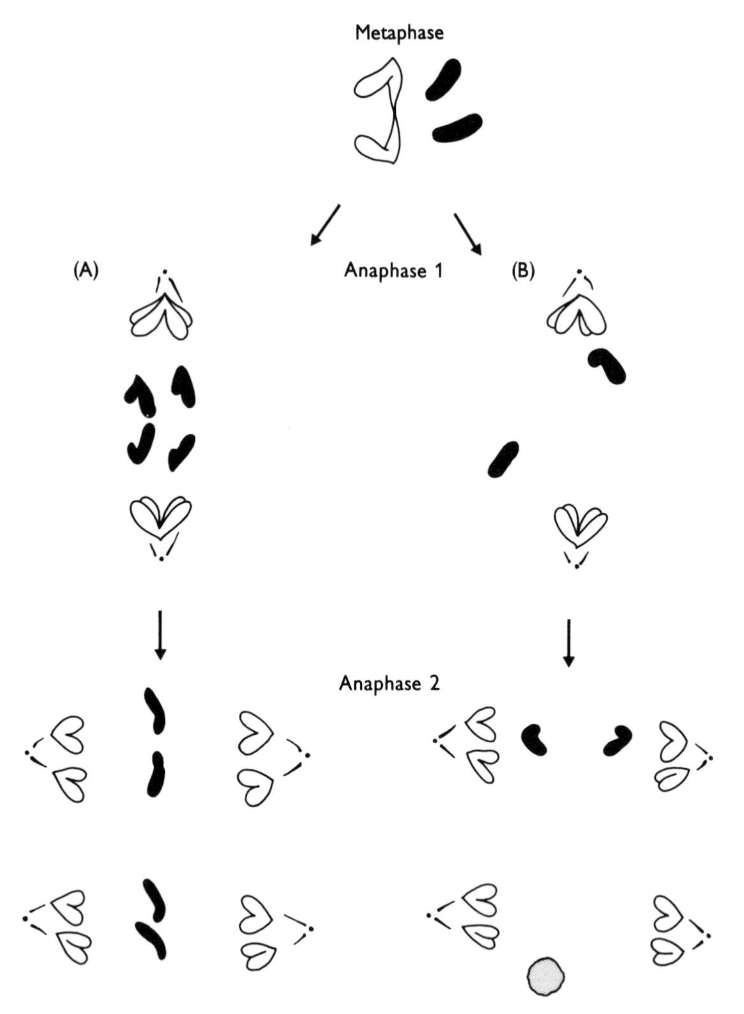

Fig. 3.2 The fate of univalents (solid black). They may divide at first anaphase (A) and be left undivided on the spindle equator at second anaphase. Alternatively (B) a univalent may be stranded on the equator at first anaphase or move to the pole and be included in the daughter nucleus. At the second division the stranded univalent is isolated in a micronucleus (shaded), the other divides normally, the daughter chromatids being included in the haploid nuclei. Another possibility is misdivision of the univalent (see Fig. 3.3).

specificity of pairing between homologous chromosomes, little is known of the forces or structures involved. More details of the structure of paired chromosomes at pachytene are revealed by the electron microscope (Moses, 1956; Westergaard and von Wettstein, 1972). Each chromosome pair is associated with a *synaptinemal complex* with the following

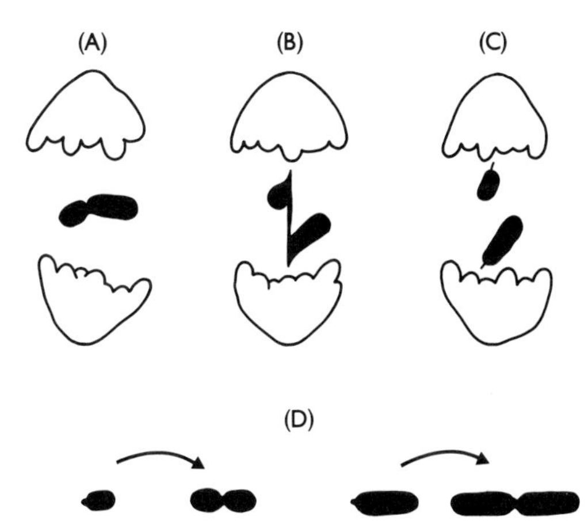

Fig. 3.3 The misdivision of univalents. The univalent (solid black) on the spindle equator at the first meiotic division (A) may undergo transverse splitting at the centromere (B) to produce two telocentric chromosomes (C). Telocentrics (D) are usually unstable and may be converted to isochromosomes.

features (see Fig. 3.4). Two dense *lateral elements* in parallel are separated by a central space which includes a central element. The lateral elements consist of protein from each of the pair of homologous chromosomes. The DNA containing chromosome fibres extend outwards in loops from these elements. The central element, like the laterals, again consists of protein emanating as loops directed inwards from the lateral elements and compressed together in the centre. The synaptinemal complex is attached at each end to the nuclear membrane by the lateral elements. The thickness of each lateral element ranges from 30–65 nm, the central space from 65–120 nm, and the central element, from 12–50 nm. The assembly of the synaptinemal complex is completed only when homologous chromosome segments are closely paired at zygotene and pachytene. During diplotene and diakinesis the synaptinemal complex disintegrates, losing first the central element and fibre loops and subsequently the lateral elements. According to Westergaard and von Wettstein (1972) the most persistent segments along the complex correspond to those at which chiasmata have formed. The implication is that the synaptinemal complex is involved in chromosome pairing and, perhaps directly, in chiasma formation. To support this view, no synaptinemal complexes are found at meiosis in male *Drosophila* where there is no crossing over or chiasma formation. The same is true of *Drosophila* females homozygous for the mutant gene *c(3)G* which suppresses crossing over. In some species carrying *X* and *Y* sex chromosomes the synaptinemal complex forms only along the paired, homologous segments of the chromosomes, at which chiasmata are formed, but not along the unpaired, differential segments, where only lateral elements develop. More difficult to accommodate are observations in haploids of maize, tomato and barley

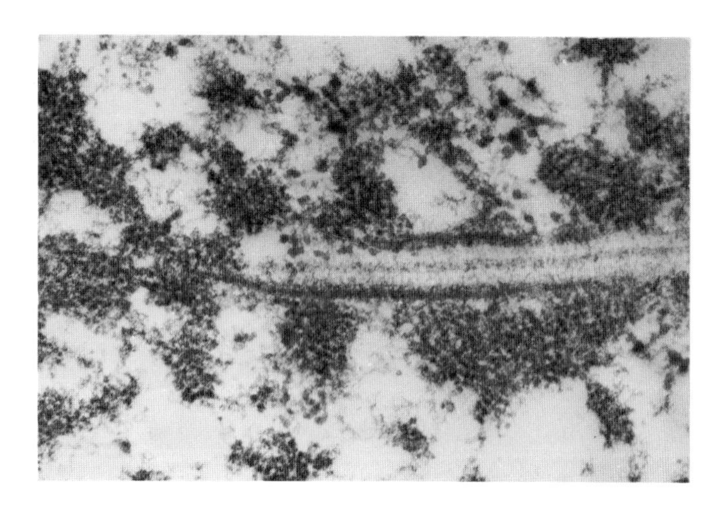

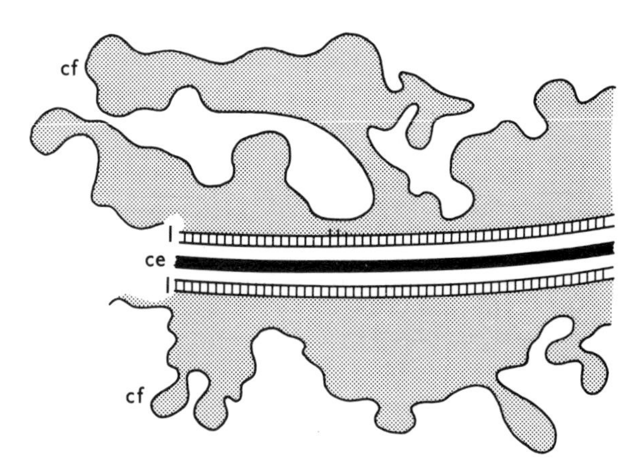

Fig. 3.4 A small section of a synaptinemal complex from a pachytene nucleus of *Lilium tigrinum*, × 45 000. ce, central element; l, lateral element; cf, chromosome fibres. [Courtesy L. F. La Cour]

which form synaptinemal complexes but where there is no chiasma formation and, apart from some degree of duplication, little homology between chromosomes. While in general therefore there are grounds for synaptinemal complexes being implicated in pairing and chiasma formation there are no clear indications of the mechanisms by which these events are accomplished.

3.4 Chiasmata

At diplotene the structure of chiasmata is readily resolved under the light microscope. The spermatocytes of grasshoppers are exceptionally favourable for this purpose. We observe that two of the four chromatids are involved in the formation of each chiasma. We infer that the chiasma results from breakage of the two chromatids followed by the rejoining of the broken ends such that there is crossing over and recombination (Fig. 3.5).

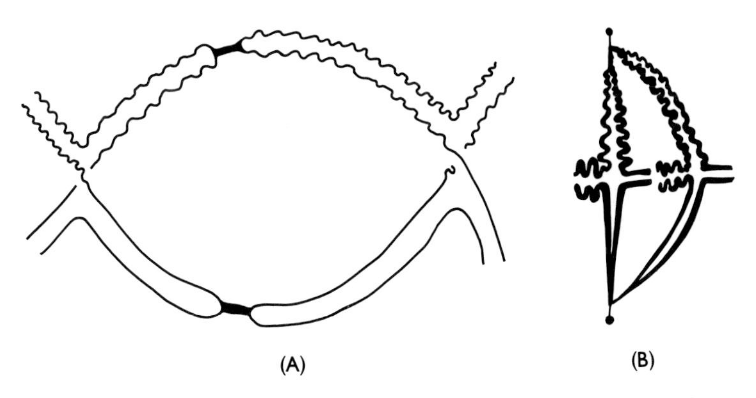

Fig. 3.5 Breakage of corresponding loci (A) followed by the re-union of broken ends (crossing over) in each arm to produce a ring bivalent (B) at first metaphase. For convenience the homologous chromosomes are distinguished by wavy and solid lines.

Co-incidence of chiasmata and crossing over

The co-incidence of chiasmata and crossing over is established from three kinds of evidence.

(a) *Cytological*

In *Lilium formosanum* Brown and Zohary (1955) found that one of the chromosomes was much shorter than its homologous partner. About two-thirds of one arm had been lost by deletion. As a result the bivalent formed is asymmetrical (Fig. 3.6). Proof of crossing over is as follows. When no chiasma forms between the unequal arms the inequality between homologous chromosomes is maintained at first anaphase. The separation is *reductional.* In contrast a chiasma in this arm leads to *equational* separation, with one arm of each of the homologous chromosomes (half-bivalents) comprising a long and short chromatid as a result of crossing over. The correspondence between the frequency with which a chiasma formed in the unequal arm and the frequency at anaphase of equational separation was extremely close, 70 and 71% respectively. Further confirmation of the correspondence of chiasma formation and of

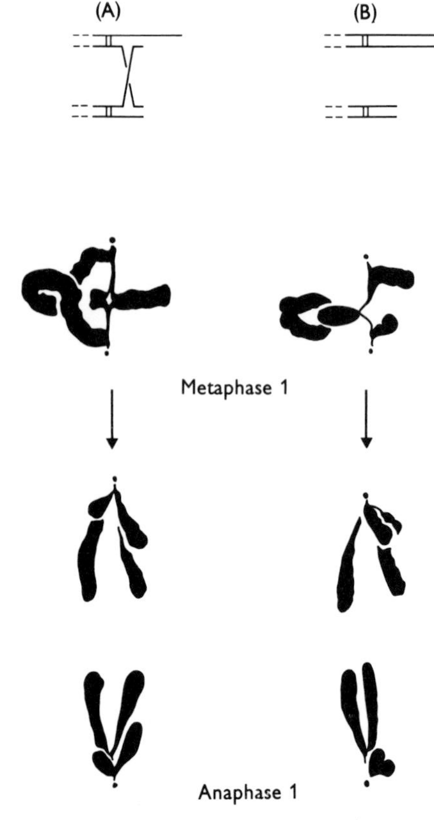

Fig. 3.6 Bivalents with one chiasma (A) and with no chiasma (B) between unequal arms of homologous chromosomes in *Lilium formosanum*. With one chiasma separation at first anaphase is equational, with no chiasma it is reductional. [After Brown, S. W. and Zohary, D. (1955) *Genetics* **40**, 850–73, with permission]

equational separation came from subjecting plants to conditions where the chiasma frequency in unequal arms was reduced to 51%. The frequency of equational separation in this case dropped, as expected, to a corresponding degree, to 50%.

(b) *Autoradiography*

In the grasshopper species *Stethophyma grossum* each bivalent has a single chiasma near to the centromere which is located terminally. By injecting tritiated thymidine into the abdomen of young males during DNA synthesis of the spermatogonial (mitotic) *S*-phase preceding meiosis, each chromosome at the commencement of meiosis has one labelled and one unlabelled chromatid (Fig. 3.7). At first anaphase of meiosis the

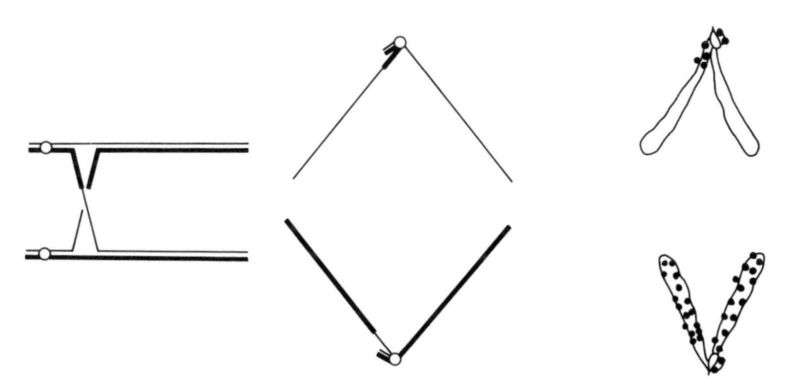

Fig. 3.7 Chiasma formation near the centromere between homologous chromosomes of *Stethophyma grossum*. One chromatid of each chromosome is labelled with tritiated thymidine (thick line), the other (thin line) is unlabelled. At first anaphase of meiosis there is a switch of label within chromatids corresponding in position to the cross-overs. [After Jones, G. H. (1971) *Chromosoma*, **34**, 367–82]

positions of exchanges between labelled and unlabelled chromatids correspond with those at which chiasmata are formed, i.e. near to the centromeres (Jones, 1971).

(c) *Linkage*

On the face of it the most convincing source of evidence that chiasmata are the results of crossing over would be to measure the amount of crossing over between marker genes within an identifiable chromosome segment and to compare this estimate of linkage with the chiasma frequency within that segment. Strange to say there is no such evidence to hand. In a more general way, however, it has been established (Table 3.2) that there is a reasonably good correlation between the chiasma frequency per bivalent at first metaphase in pollen mother cells of maize and the amount of crossing over as estimated by linkage (Whitehouse, 1969).

Chiasmata, crossing over and recombination

Crossing over, as we have indicated, is assumed to take place at the site of chiasma formation. At each chiasma two of the four chromatids will be cross-overs, the others unchanged so that two of the four gametes resulting will incorporate cross-overs, the remaining two non-cross-overs at that site. The relation between chiasma frequency and cross-over frequency is, therefore, 2:1. With suitable gene markers on each side of the chiasma and near to it, all the cross-overs will be detectable as recombinants. If, however, the markers are so far apart as to allow for more than one chiasma between them the cross-overs will, on occasion, be undetectable as recombinants. While the relation therefore between

Table 3.2 Comparisons of chiasma and cross-over frequencies in *Zea Mays.* The chiasma frequency per chromosome is derived from the mean chiasma frequency per cell, assuming that the chiasma frequency is proportional to chromosome length. The cross-over frequency is based on linkage data for the gene markers available in each chromosome. These may not embrace the entire length of each chromosome. [Data from Whitehouse, H. L. K. (1969) *Towards an Understanding of the Mechanism of Heredity*, 2nd ed., Edward Arnold, London]

Chromosome	Length in μm at mid-pachytene	Estimated mean chiasma frequency	Mean cross-over frequency
1	82	4.0	3.1
2	67	3.3	2.6
3	62	3.0	2.4
4	59	2.9	2.2
5	60	2.9	1.4
6	49	2.4	1.3
7	47	2.3	1.9
8	47	2.3	0.6
9	43	2.1	1.4
10	37	1.8	1.1
Totals	553	27.0	18.0

chiasma frequency and crossing over frequency is a constant at 2:1, the recombination value, as an index of crossing over, becomes less reliable with increasing distance between markers. However frequent the chiasma formation and crossing over within a bivalent the maximum recombination between markers will be 50%. In contrast, the frequency of chiasmata and of crossing over between markers may be, and often is, much higher. The situation with respect to two chiasmata between widely spaced markers is illustrated in Fig. 3.8.

The time of chiasma formation

In eukaryotes the evidence points firmly to the formation of chiasmata at early prophase of meiosis. That chiasmata are formed after the synthesis of DNA at the premeiotic interphase is clear from the work of Rossen and Westergaard (1966). In the Ascomycete *Neottiella rutilans,* DNA synthesis takes place in haploid nuclei before they fuse to produce the diploid nucleus which undergoes meiosis during which the chiasmata are formed.

Figure 3.9 shows the effects of heat treatment upon chiasma frequencies in the spermatocytes of *Schistocerca gregaria* (Henderson, 1970). It will be observed that treatments applied up to the zygotene/pachytene stage affect the chiasma frequencies. The conclusion is that chiasmata are formed at zygotene/pachytene but not later than this stage. There is no evidence that they are formed earlier. It is pertinent to recall that it is at

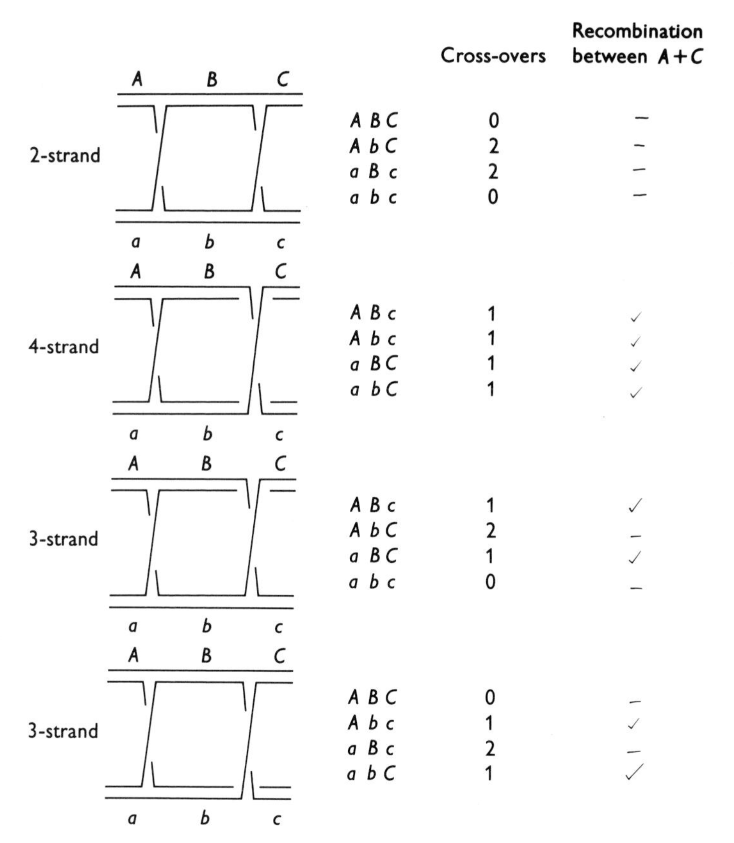

Fig. 3.8 The four possibilities when chiasmata form simultaneously between *A* and *B* and *B* and *C*. The four possibilities are realized with equal frequency showing there is no 'chromatid interference'. On the right are the recombinant gametes, the number of cross-overs between *A* and *C* and, far right, the number of recombinants for the outside markers, *A* and *C*. The ratio between the chiasma frequency and the cross-over frequency is a consistent 1:2. For any pair of markers, e.g. the outside markers A and C, the recombination overall, is 50%. In the absence of chromatid interference this is the maximum. The diagram shows that for the construction of linkage maps which are based on the cross-over frequencies it is important to choose markers as close together as possible. If, as with *A* and *C*, the distance between them allows for the formation of two or more chiasmata simultaneously the double cross-overs, A*bC* and *aBc* would be undetected.

this zygotene/pachytene stage that the synaptinemal complexes are fully assembled; also that they are indispensable prerequisites for chiasma formation in eukaryotes. There is little doubt therefore that chiasma formation is restricted to the zygotene/pachytene stage. This would suggest that the intimate association of homologous chromosomes

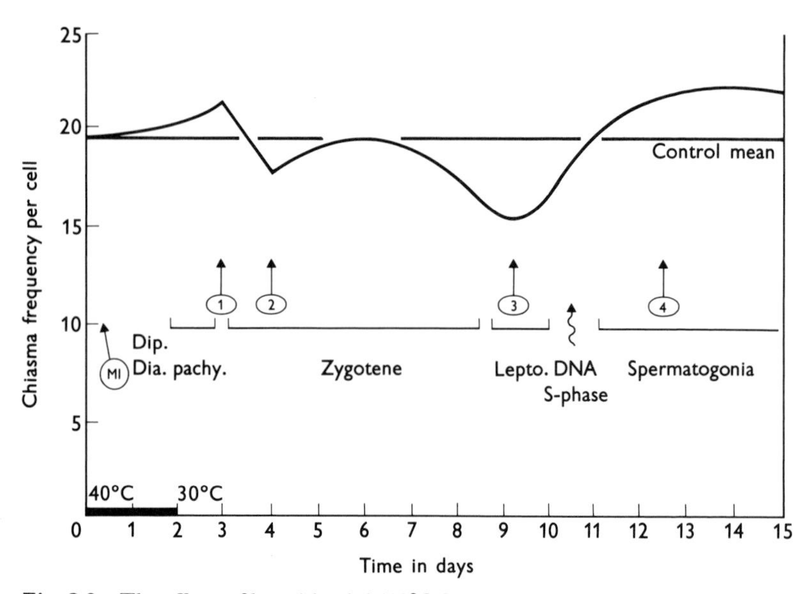

Fig. 3.9 The effects of heat 'shocks' (40°C for 2 days) upon chiasma frequencies at first metaphase of meiosis in spermatocytes of *Schistocerca gregaria*. The shocks are effective up to late zygotene (2) and early pachytene (1) but at no later stage, suggesting that chiasma formation is completed in pachytene. The decrease and increase respectively in chiasma frequencies due to shocks at leptotene (3) and the interphase preceding DNA synthesis (4) reflect disturbances of processes which affect the formation of chiasmata at the later stage. The time intervals between treatments and the manifestation of their effects at first metaphase are established by reference to the horizontal scale. [From Henderson, S. A. (1970) *Ann. Rev. Genet.*, **4**, 295–324]

initiated at zygotene is necessary for the formation of chiasmata. There are numerous observations to support this view. In maize the amount of crossing over, and therefore the frequency of chiasmata, is directly proportional to the completeness of chromosome pairing at pachytene. In the pollen mother cells of a heterozygote carrying a normal chromosome 9 and an abnormal 9 which includes a transposed segment from chromosome 3, the amount of intimate pairing at pachytene between these homologous chromosomes is reduced, and so also is the amount of crossing over (Rhoades, 1968).

Figure 3.9 shows that heat treatments applied at leptotene, and even before the synthesis of DNA, affect chiasma formation. Heat shocks (or low doses of radiation) have similar effects in *Tradescantia* (Lawrence, 1961), *Drosophila* (Grell and Chandley, 1965) and the grasshopper *Gonia australasiae* (Peacock, 1970). They are no doubt symptoms of disturbances of processes which directly or indirectly affect the formation of chiasmata at a later stage. Precisely what these processes are is not known.

The question of interlocking

At first metaphase of meiosis bivalents are, on occasions, interlocked. The cause of the interlocking is obvious enough (Fig. 3.10). What is not obvious is why interlocking is such a rare event because the chromosomes at zygotene and pachytene are still uncoiled and extended. It is certain therefore that before the intimate pairing of homologous chromosomes that is required for chiasma formation there must be some preliminary separation and alignment of homologues to prevent interlocking when the chiasmata are formed. One suggestion is that such alignment takes place at the pre-meiotic mitosis such that homologues are already contiguous at the start of the interphase of meiosis. The evidence for this is conflicting. Moreover in many haploids, such as the Ascomycetes among fungi, the homologous chromosomes are not brought together until the fusion of gametic nuclei immediately prior to meiosis itself. There are no reports of widespread interlocking in these or comparable species. We would suggest that the alignment takes place during interphase itself when, as is established from cinematographs, the chromosomes are in vigorous movement. The alignment could be achieved as follows: (1) at interphase the homologous chromosomes are associated at or near one end (telomere); (2) the paired telomeres move through the nucleus pulling the chromosomes for a distance equal to or greater than the length of the chromosomes. By this means each pair, although associated closely only at one point, is now disentangled from the remainder.

It is well established that the paired telomeres at both ends of homologous pairs of chromosomes become attached to the nuclear membrane well before the onset of pachytene (Moens, 1969a). By this means the separation of homologous pairs achieved by movement will have been consolidated and the possibility of interlocking reduced to a minimum

Fig. 3.10 Pairing at pachytene leading to the interlocking of bivalents at first metaphase of meiosis.

by the time the chromosomes begin their intimate association around the synaptinemal complexes.

The occasional and exceptional cases of interlocking such as may be induced by, for example, heat shock are explained by restriction of movement during interphase, or by the failure of telomere association or of their attachment to the nuclear membrane.

Chiasma distribution

(a) *The question of chromatid interference*

Where two or more chiasmata form within the one chromosome are the two chromatids involved in crossing over at any one chiasma independent of those at other chiasmata? Put in another way, is there *interference* with respect to chromatids such that a chromatid involved at one chiasma is more, or less, likely to be involved at another. In Fig. 3.8 we have, in respect of each pair of chiasmata, assumed no chromatid interference such that each of the four alternative types apply with equal frequency. The assumption is justified by evidence from linkage experiments. They show that the possibilities in Fig. 3.8 occur at random (but see also p. 52).

(b) *Chiasma localization*

In general more chiasmata are found in the longer than in the shorter chromosomes within a complement. The relation between chiasma frequency and chromosome length, however, is not linear. Mather's analysis of meiosis in pollen mother cells of numerous plant species shows that the chiasma frequency per unit length is characteristically higher for short than for long chromosomes (Fig. 3.11). The figure shows that at

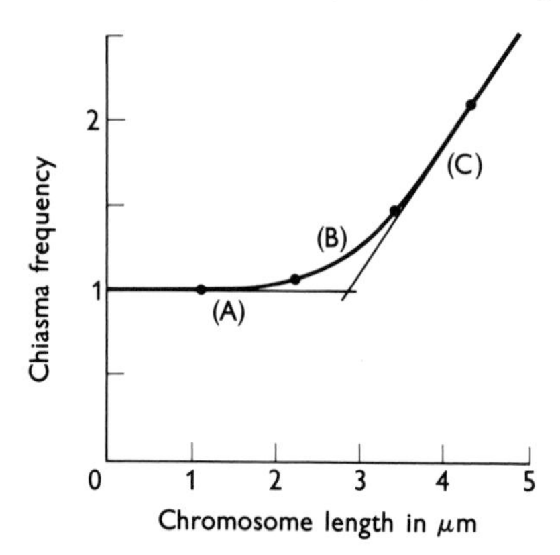

Fig. 3.11 The relationship between chiasma frequency and chromosome length. For very short chromosomes (A) there is a chiasma frequency of 1 irrespective of chromosome length. At (B) the relationship changes such that at (C), the chiasma frequency increases with increasing chromosome length. [From Mather, K. (1938) *Biol. Rev.*, **13**, 252–92, with permission]

least one chiasma is formed even in very short chromosomes. From an adaptive standpoint one readily appreciates that if these short chromosomes were unable to form a chiasma regularly at meiosis they could not persist. Chiasma failure results in univalents and, as we have explained, univalents are, more often than not, 'lost' during meiosis.

From the standpoint of mechanics the factors which determine the distribution of chiasmata are by no means clear. On a descriptive basis, however, the model proposed by Mather in 1937 is especially useful. The model envisages that chiasma formation is sequential. The first chiasma forms at a distance *d*, the *differential distance*, from an initiation point which coincides with the point of initial pairing at zygotene. Since even short chromosomes form at least one chiasma this distance may be short, even approaching zero (Henderson, 1963). Cytological observations show that subsequent chiasmata form at fairly regular intervals apart. The inference is that there is interference between chiasmata such that they do not cluster. The inference is verified from linkage data which show chiasma interference: *i* in Mather's notation. For all bivalents within a chromosome complement, therefore, the location of chiasmata follows a well-defined pattern. Moreover the frequency per bivalent will relate directly to *d* and *i*. For a short chromosome, where the first chiasma forms at a distance *d* from, say the end of one arm, a second chiasma is precluded if the remaining segment is shorter than *i*. For longer chromosomes the number of chiasmata formed subsequent to the first will be equal to the distance distal to the first chiasma, divided by *i*. The pattern, though well defined, is not rigid. Both *d* and *i* are subject to fluctuation. Even so the model makes abundantly clear that chiasma distribution is not random. A most important implication is that chiasmata and crossovers are more numerous in some chromosome segments than others. Put in another way, it means that some gene sequences are, to a degree, 'protected' from disruption by crossing over, others are not. The maintenance of relatively intact sequences of genes may, from an adaptive standpoint, be as important in conserving genetic variability as the disruption by recombination within other sequences to generate new variability.

There are, of course, numerous variations on the standard sequence of events which take place at meiosis including those affecting chiasma formation (see John and Lewis, 1965). It is not unusual either for the female meiosis to be substantially different from the male. A familiar example is the absence of crossing over at meiosis in *Drosophila* males in comparison with the normal chiasmate meiosis in the female. There are instances also where the *distribution* of chiasmata differs sharply between male and female. In *Stethophyma grossum* males for example the chiasmata are strictly localized near to the centromeres; in the female they are widely distributed (Perry and Jones, 1974). Extreme localization of chiasmata as in the male *Stethophyma* may stem from more than one cause. Two flowering plant species, *Fritillaria meleagris* and *Allium fistulosum*, have chiasmata localized mainly near to the centromeres. In *F. meleagris* the pachytene pairing is complete only in the vicinity of the centromeres. Consequently the chiasmata are confined to the centromere regions (Darlington, 1935). The same explanation probably applies also in cases where crossing over is clustered within very short chromosome segments during meiosis in some fungi (Fincham and Day, 1971). In *Allium fistulosum*, on the other hand, the explanation does not hold as pairing at pachytene is complete.

3.5 The mechanism of recombination

Reduction of the chromosome number at meiosis is accounted for by bivalent formation at first prophase followed by the separation of half bivalents at first anaphase and of chromatids at the second anaphase. Recombination between genes on different chromosomes is explained by the random orientation of bivalents at first metaphase such that genes on different bivalents segregate independently to yield 50% recombinant and 50% parental type gametes (Fig. 3.12). The recombination of genes on the same chromosome has been interpreted on the basis of breakage of non-sister chromatids at precisely corresponding places followed by the reunion of the broken ends resulting in crossing over (Fig. 3.5). The mechanism depicted also accounts for the reciprocal nature of the products of recombination and for the numerical equality of reciprocal products characteristic of results from the vast majority of linkage analyses. In micro-organisms however, where the precision of linkage analysis is enhanced enormously on account of the large number of progenies available for classification and, in the *Ascomycetes*, further refined by the opportunity of classifying the products of individual meiotic divisions (Fig. 3.13), the results have brought to light instances where the products of recombination are not reciprocal. Any model which seeks to accommodate the facts of recombination at meiosis must take account of these non-reciprocal recombinants. The breakage–reunion model in Fig. 3.5 does not.

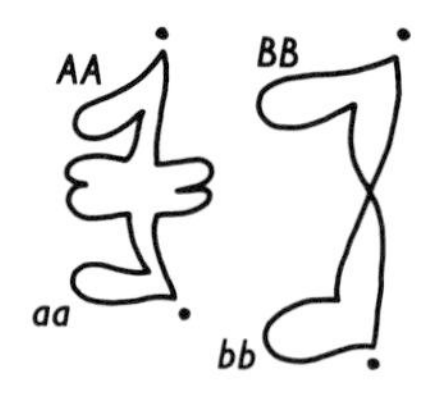

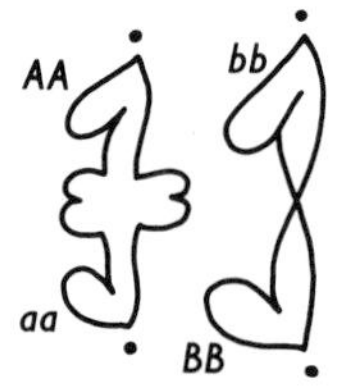

Gametes *AB:AB:ab:ab* *Ab:Ab:aB:aB*

Fig. 3.12 Random orientation of two bivalents at first metaphase of meiosis, giving independent segregation of genes *A* and *B*.

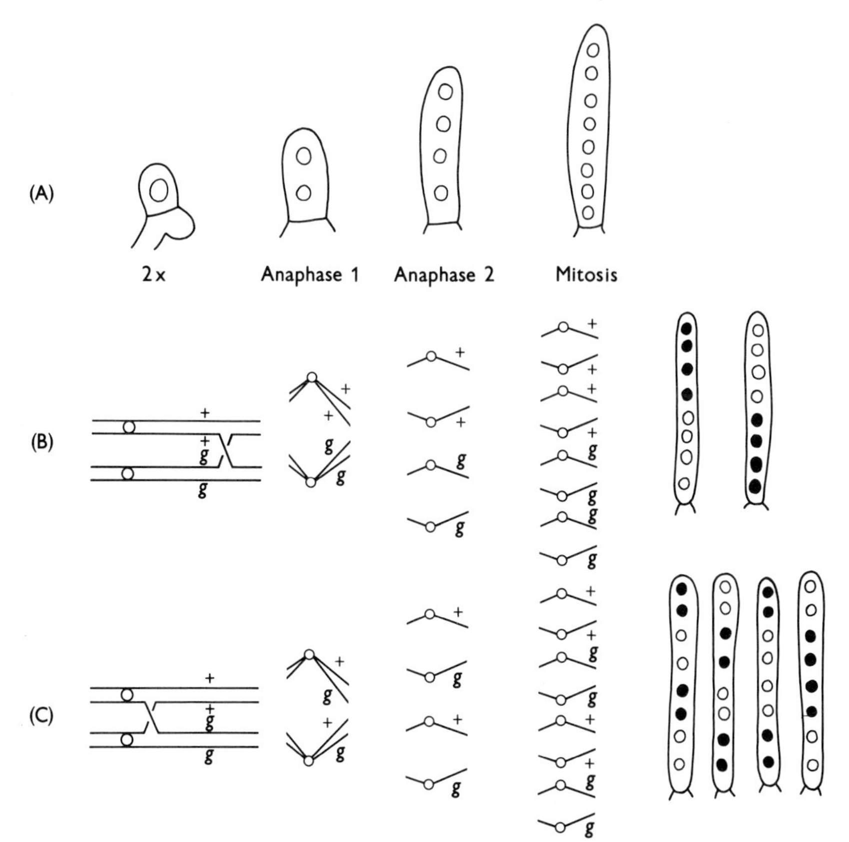

Fig. 3.13 Tetrad analysis in *Sordaria fimicola*. Each mature ascus contains the products of one meiosis and the ascospores are ordered in relation to the first and second division (A). Spore colour markers show first division segregation (B) in the absence of chiasma formation between the marker gene and the centromere and second division segregation (C) when a chiasma is formed.

Gene conversion

Crosses between a haploid strain of *Sordaria fimicola* with wild type black (+) ascospores and a strain with mutant grey (*g*) ascospores produce asci containing two basic types of spore arrangement, the *first division segregation* (reductional) type, where there is no crossing over between the spore colour locus and the centromere and the *second division segregation* (equational) type, where crossing over has occurred in this region (Fig. 3.13). In every ascus, irrespective of whether crossing over has taken place or not, we expect 4 black and 4 grey ascospores, i.e. a 1 : 1 segregation. This, in fact, is what we do find in the vast majority of asci examined. In others, however, instead of the exclusive occurrence of 4 : 4 segregations for black and grey, some 6 : 2 and 5 : 3 segregation patterns

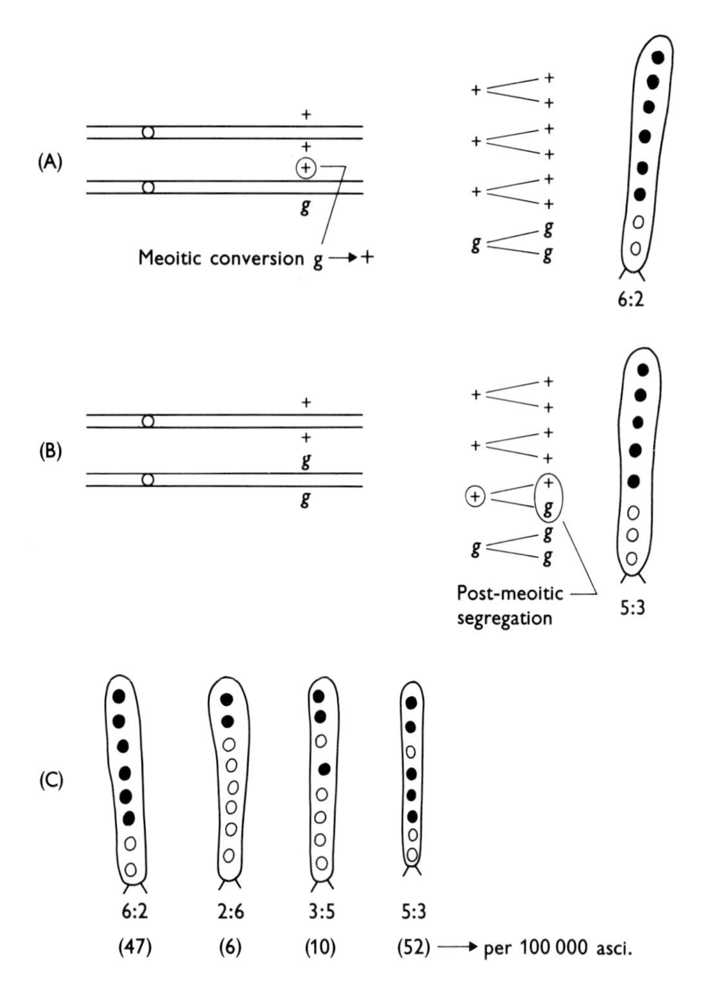

Fig. 3.14 Aberrant asci in *Sordaria fimicola*. (A) Conversion of the *g* allele at meiosis is detected as a 6 : 2 segregation of black to mutant ascospores. (B) 5 : 3 asci arise from the post-meiotic segregation of heterozygosity within a single chromatid. The frequencies of 6 : 2 and 5 : 3 conversion asci for the *g* locus of *S. fimicola* are given in (C). [Data from Kitani, Y., Olive, L. S. and El-Ani, A. S. (1962) *Am. J. Bot.*, **49**, 697–706]

turn up with low frequency (Fig. 3.14). The 5:3 segregations are particularly interesting because they represent post-meiotic segregation of differences within a single chromatid. Mutation will not account for these *aberrant* asci because, first, they are too numerous and, second, the change is always in the direction of one of the existing alleles and not at random. The alternative is *gene conversion*—'the replacement of the genetic material at a particular site in one chromatid by the genetic material from a precisely corresponding site from a non-sister chromatid' (Fincham and Day, 1971). The phenomenon accounts for the non-reciprocal recombinants to which we have referred (Fig. 3.14).

(a) *Reciprocal and non-reciprocal intragenic recombination*

Lissouba *et al.* (1962) collected several thousand pale-coloured ascospore mutants in *Ascobolus immersus.* On the basis of linkage the mutants were assigned to a number of different *series.* Within each series the mutants were tightly linked and each series was considered to represent one gene locus. Crosses between pale-spored mutants located in different genes yielded some wild type and double mutant reciprocal recombinant classes as well as the two single mutant parental types. Crosses between mutants located in the same gene (i.e. allelic mutants) also yielded some wild type recombinants but test crosses with the parental strains showed that this intragenic recombination was not reciprocal (Fig. 3.15). In *Ascobolus* there was also an exception to the general rule. Crosses between some of the mutants at one locus (in series 73) produced a high propor-

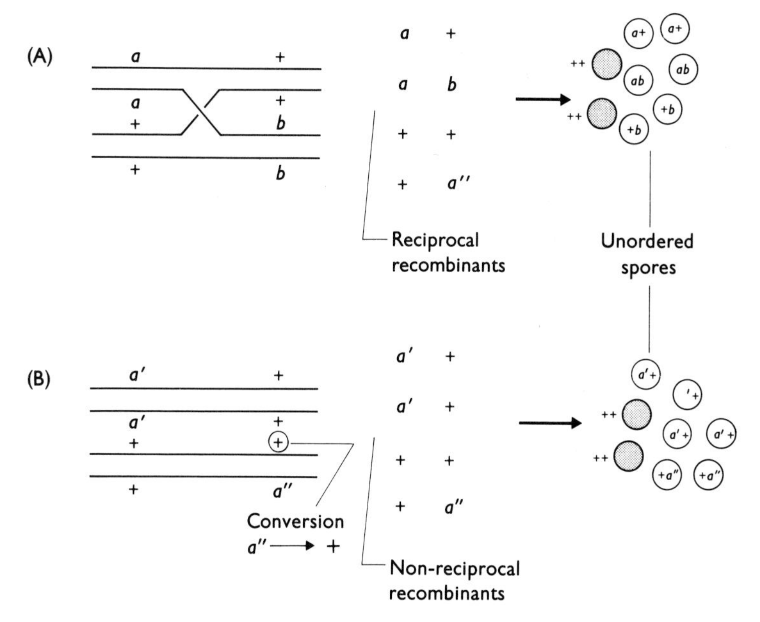

Fig. 3.15 Intergenic and intragenic recombination in *Ascobolus immersus.* (A) Crosses between mutants of different genes (*a* and *b*) yield 6:2 asci arising by reciprocal crossing over. (B) Pairwise crosses between mutants of the same gene (a' and a'') also give 6:2 asci, but these are found to contain the products of non-reciprocal recombination, arising by conversion. Note the spores in this species are not ordered as in *Sordaria*.

tion of reciprocal recombinants. Reciprocal intragenic recombination has also been detected for the *arg-4* gene of *Saccharomyces cerevisiae* (Fogel and Mortimer, 1969), so that intragenic recombinants are not exclusively the result of conversion.

(b) *Polarity*

When a number of *Ascobolus* mutants from within the one gene are crossed in all possible combinations there is a well defined hierarchy or polarity with regard to conversion such that one mutant, e.g. *a* in a series *a b c d* is converted in crosses with *b*, *c* or *d*; *b* when crossed with *c* or *d* and *c* when crossed with *d*. Moreover this polarity is directly correlated in *Ascobolus*, as in other but not all *Ascomytes*, with the spatial distribution of the mutant sites, i.e. their linkage relations. The polarity may, in fact, be used for mapping the mutants within the gene (Lewis and John, 1970).

(c) *Negative interference and chromatid interference*

In asci showing conversion (e.g. at the *g* locus in *Sordaria*, Kitani *et al.*, 1962) there is an unexpectedly high frequency of recombination between the gene undergoing conversion and closely linked markers on either side. Double recombinants are also unexpectedly numerous. In view of the short distances involved the double recombination is suggestive of negative interference between chiasmata and, moreover, of chromatid interference during chiasma formation, since the double recombination implicates the same pair of chromatids. As we shall explain the excess of double recombinants may be accounted for without invoking interference between chiasmata.

It is also important to emphasize that when the region is mapped on the basis of a random sample of asci, as distinct from those showing conversion, there is no detectable negative interference or chromatid interference. The reason is that conversion at the *g* locus (or any other) occurs with very low frequency, in one out of about 1000 asci. It is for the same reason that non-reciprocal recombinants are undetectable by conventional linkage analysis. Even so the three phenomena, the high recombination frequency within short chromosome segments, the excess of double recombinants and the non-reciprocal recombinants are real enough. In constructing a general model which accounts for the mechanism of recombination they must be taken into account. A number of such models have been proposed. We present below that put forward by Holliday.

The molecular basis—Holliday's model (1964)

1 Each chromatid is composed of one double helix of DNA.
2 Recombination takes place at the four strand stage after chromosome pairing and duplication.
3 The DNA from non-sister chromatids is closely paired (Fig. 3.16A).
4 Simultaneous breakage at strictly corresponding sites in two single

chains (half-helices) of the same molecular polarity is followed, first, by unwinding of the half-helices in one direction for a variable length and, second, by the union of ends in new combinations to form a 'half-chiasma' and a region of hybrid DNA (Fig. 3.16B).

5 Either the half-chiasma is resolved by iso-locus breakage in the overlapping strands, the result being no recombination for outside markers (Fig. 3.16C), or alternatively, the half-chiasma is converted into a full chiasma by iso-locus breaks in the non-overlapping strands, which leads to reciprocal recombination for the outside markers (Fig. 3.16D).

6 There is correction of 'illegitimate' pairing of bases for the mutant site spanned by the region of hybrid DNA. The result is either a 4:4 or a 6:2 ascus-type segregation, depending on the way in which the correction takes place (Fig. 3.16E 1, 2). Delay in the correction of one of the hybrid DNA molecules until the post-meiotic DNA synthesis period accounts for the 5:3 post-meiotic segregations (Fig. 3.16E 3).

The model shows that: (a) gene conversion is achieved with or without recombination for outside markers, depending upon whether a half-chiasma or a full chiasma is formed. Conversion will, of course, only take place at all when the hybrid DNA is formed within a site of heterozygosity. (b) The 'negative interference' is the outcome of conversion and of the formation of a chiasma as in Fig. 3.16C. (c) The unwinding of half-helices from a fixed point in one direction accounts for the polarity of conversion. Mutant sites nearest to the breakage point are the most likely, therefore, to undergo conversion, and vice versa. (d) Reciprocal recombinants between mutant sites within a gene are produced if both

Fig. 3.16 Holliday's 1964 model for the mechanism of recombination. The model (A) represents paired homologues at the four-strand stage and includes a single gene, with two sites of heterozygosity, two outside marker genes (*A*, *B*) and the centromeres. Each chromatid consists of a double helix of DNA drawn as a pair of straight lines. The arrows indicate the polarities of the DNA half-helices. Recombination is initiated by simultaneous breakage at strictly corresponding sites (arrowed) in each of two single chains of the same polarity in DNA helices from non-sister chromatids. The half-helices unwind in one direction for a variable length and then rejoin in a new combination to form a half-chiasma (B). The half-chiasma may either be resolved by an iso-locus break in the crossed over strands (C), when there is no recombination for the outside markers, or else become converted into a full chiasma (D) following iso-locus breaks in the non-crossed over strands, when outside marker recombination will occur (E). In either event the region of hybrid DNA spans a site of heterozygosity and thus includes the mismatched base pairs A:C/T:G. Correction of this illegitimate pairing before replication can lead to a normal 4:4 segregation of alleles at site M1①, a 6:2 segregation ②, or a 5:3 segregation ③, depending on the direction (in terms of parental types) in which the correction is made (F).

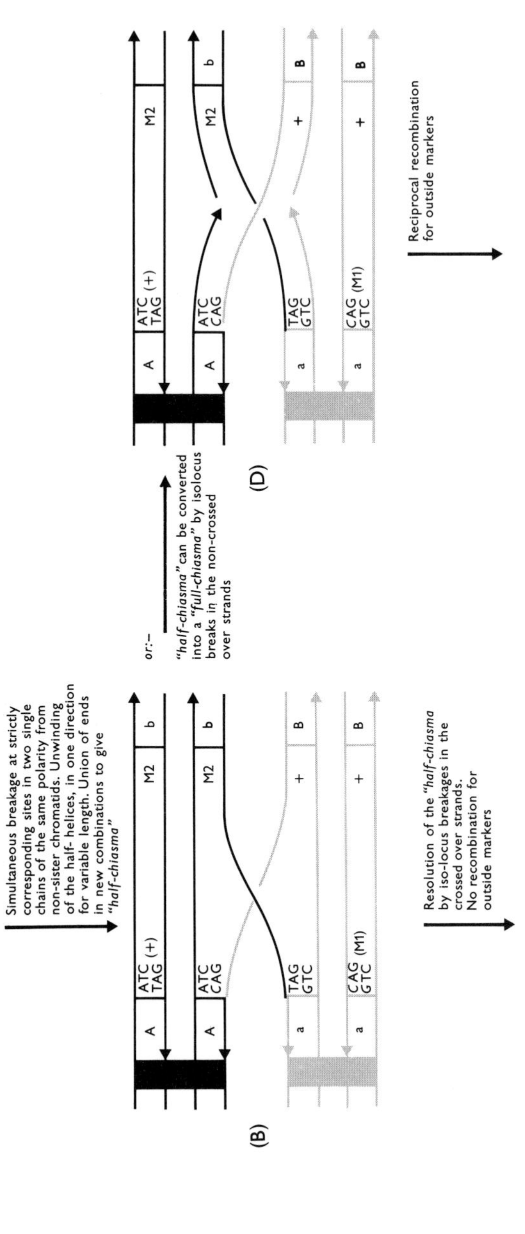
(A)
A
ATC
TAG
(+)
M2
b
a
CAG
GTC
(M1)
+
B
Simultaneous breakage at strictly corresponding sites in two single chains of the same polarity from non-sister chromatids. Unwinding of the half- helices, in one direction for variable length. Union of ends in new combinations to give *"half-chiasma"*
(B)
ATC
CAG
TAG
GTC
Resolution of the *"half-chiasma* by iso-locus breakages in the crossed over strands. No recombination for outside markers
or:–
"half-chiasma" can be converted into a *"full-chiasma"* by isolocus breaks in the non-crossed over strands
(D)
Reciprocal recombination for outside markers

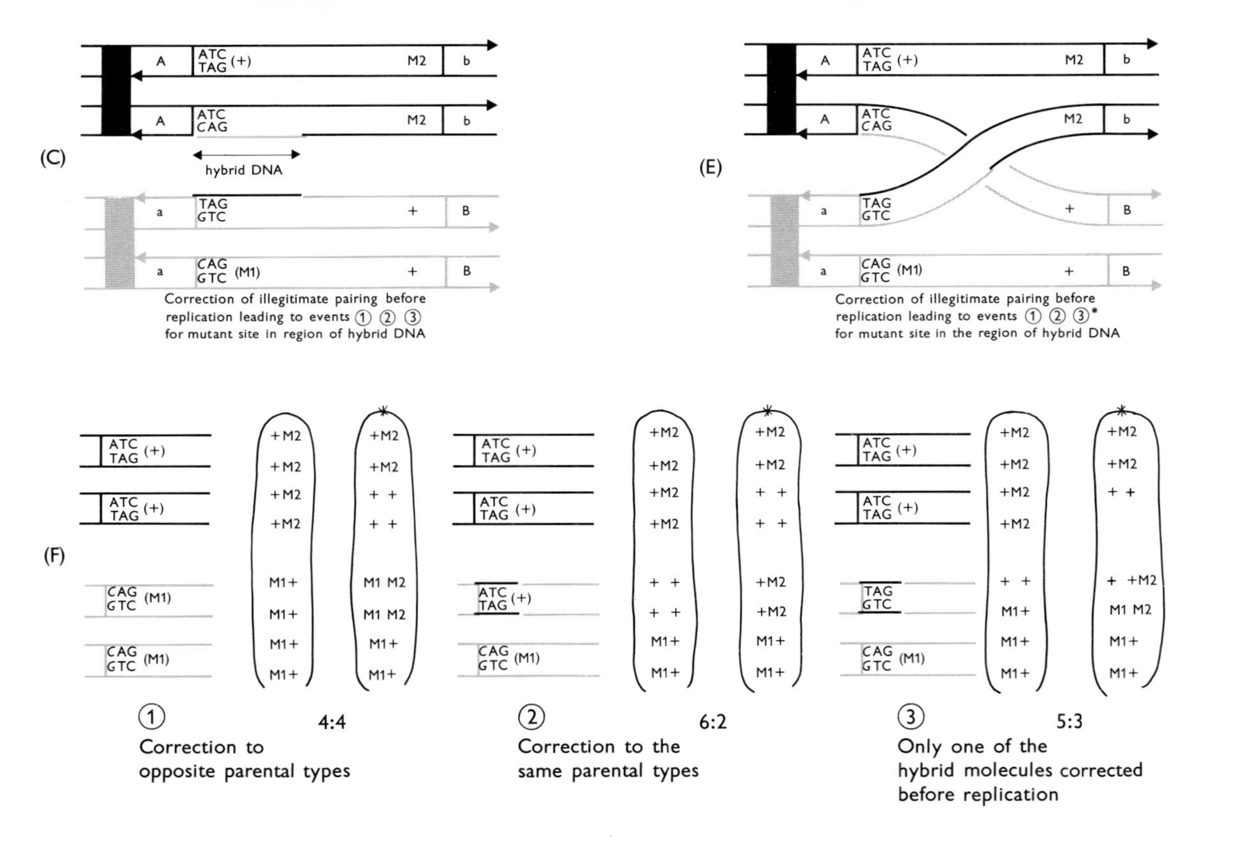
(C)
A
ATC
TAG
(+)
M2
b
ATC
CAG
hybrid DNA
a
TAG
GTC
+
B
CAG
GTC
(M1)
Correction of illegitimate pairing before replication leading to events ① ② ③ for mutant site in region of hybrid DNA
(E)
Correction of illegitimate pairing before replication leading to events ① ② ③* for mutant site in the region of hybrid DNA
(F)
+M2
M1+
M1 M2
+ +
① Correction to opposite parental types
4:4
② Correction to the same parental types
6:2
③ Only one of the hybrid molecules corrected before replication
5:3

mutants are to one side of the hybrid DNA or on either side but outside the hybrid DNA region.

At this point it is worth re-emphasizing that for any one of numerous loci within a chromosome the frequency with which it undergoes conversion is low. Moreover, the consequences of conversion, such as non-reciprocal recombinants affect only a restricted chromosome segment in the immediate vicinity of the conversion. For these reasons the general relations between chiasmata crossing over and recombination correspond very closely with those which apply in models which assume single breaks in non-sister chromatids followed by reunion of the broken ends as in Fig. 3.5, even though, as we have seen, the molecular basis for recombination is substantially different from this.

Models invoking the formation of hybrid DNA such as those of Holliday and of Whitehouse (1963) are the more plausible in the light of more recent discoveries. We have already referred to the small fraction of the DNA which is synthesized at zygotene/pachytene when the 'correction' of mismatched base sequences is assumed to take place. Enzyme systems have been described also, with the capacity for inducing breakage, repair and the synthesis of DNA in meiotic cells. An endonuclease, found exclusively in meiotic cells, has the property of inducing single-strand 'nicks' in native DNA. Its activity is at a peak during late zygotene and early pachytene. A ligase 'repair' enzyme also shows a flush of activity at the same period (Howell and Stern, 1971). Stern and Hotta (1973) have also shown that DNA extracted from meiotic cells of *Lilium* and denatured contains a higher frequency of short strands during pachytene than at other stages of meiosis. They ascribe the high frequency of these short strands to the endogenous nicking by the endonuclease and their subsequent disappearance to the repair mediated by the ligase.

A full account of the events associated with gene conversion and recombination is given in a companion volume in this series (Catcheside, 1977). This also deals in detail with the various molecular models which have been proposed to account for the diverse phenomena associated with recombination.

4

Quantitative variation in genetic material

4.1 The structural basis

Evolution is the result of the selection of phenotypes displaying adaptive variation in heredity. The heritable variation stems, basically, from changes in the information which resides mainly in the chromosomal DNA. The changes are of two kinds, qualitative and quantitative. At the molecular level the qualitative changes result from base substitution within the DNA molecule, which affects the *base ratio*, or from rearrangement affecting the order of bases (Fig. 4.1). Examples of both kinds are revealed by direct comparisons between DNAs of species throughout both plant and animal orders. In the bacteria the DNA base ratio (expressed as the proportion of guanine + cytosine + 5-methylcytosine where present), ranges from 25 to 75% (Belozersky and Spirin, 1960; Sueoka, 1961). The G + C content in higher plants ranges from 35 to 49%.

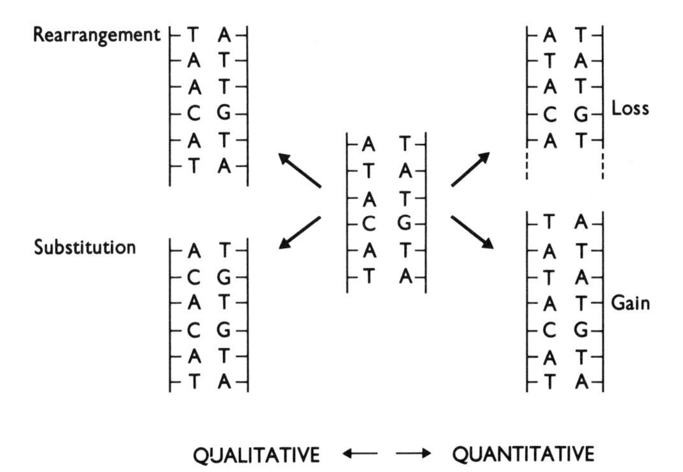

Fig. 4.1 Qualitative changes in DNA affecting the base ratio and the base sequence.

A 10% range is found even within the one family, the Liliaceae (Kirk, Rees and Evans, 1970). Changes in order are less easily established. One example, involving RNA in this case, is given by Reddi (1959) who showed differences in the order of bases within the nucleic acids of different strains of tobacco mosaic virus. Turning to *quantitative* variation we would expect to find an increase in DNA amount accompanying the evolution of more complex and sophisticated forms. The amount of chromosomal DNA in vertebrate animals and higher plants is, in fact, 1000 times greater than in bacteria (Fig. 4.2). We have commented earlier, however, that variation of this order of magnitude is difficult to explain directly in terms of increase in the variety of genes within the complement. The large variation in DNA amount between closely related species, e.g. in *Vicia* and *Lathyrus* (Rees *et al.*, 1966) is equally incomprehensible on grounds of variation in the number and variety of nuclear genes. What then is the explanation for this wide range of variation in nuclear DNA? How does it arise? What kind of DNA is involved? What does it signify in genetic terms?

In structural terms the causes of quantitative variation in nuclear DNA arise either by the addition or deletion of whole chromosomes, i.e. a change in chromosome number or, alternatively, by the addition or deletion of segments within chromosomes.

Polyploidy and aneuploidy

The addition of whole sets of chromosomes follows the failure of spindle formation at mitosis in somatic cells giving rise in a $2x$, diploid organism to $4x$, tetraploid somatic nuclei. An alternative cause of polyploidy is non-reduction during gamete formation resulting for example in the production of diploid $2x$ rather than haploid gametes at meiosis in a diploid. The fusion of $2x$ and x gametes produces $3x$, triploid zygotes, the fusion of diploid gametes, $4x$ tetraploids. With aneuploidy the gain or loss of chromosomes involves less than a whole set or genome. The gain of one chromosome of the complement gives a trisomic $(2x+1)$, of two members of the complement a tetrasomic $(2x+2)$ and so on. Monosomics $(2x-1)$ and nullisomics $(2x-2)$ are the results of the loss of one or of a pair of homologous chromosomes, respectively.

The nature of the DNA lost or gained through polyploidy and aneuploidy is clear enough. It represents exact copies of complete chromosome complements or of complete chromosomes. In the long term it is possible that small segments become redundant and are deleted from the chromosomes of tetraploids. Timmis, Sinclair and Ingle (1972) and Flavell and Smith (1974) report a loss of ribosomal cistrons in polyploid hyacinths and wheats respectively.

(a) *Autopolyploids*

These are polyploids in which the additional sets of chromosomes are structurally similar and therefore homologous to those of the diploid

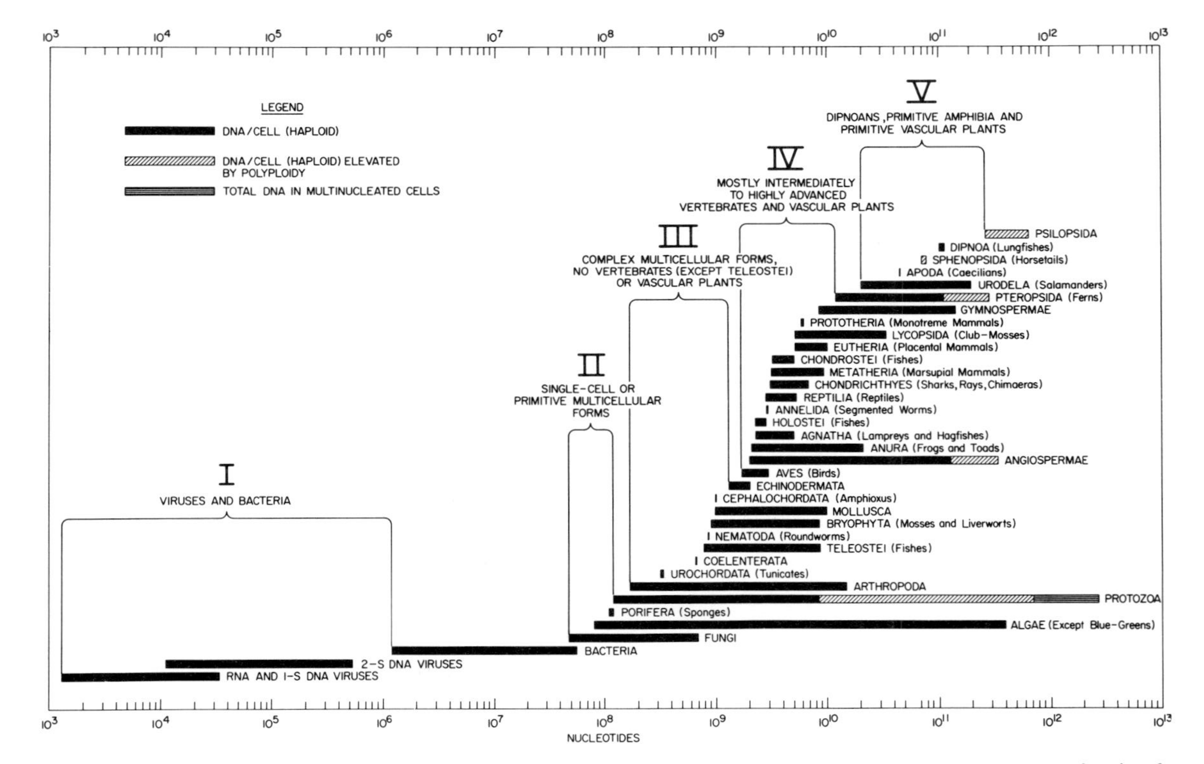

Fig. 4.2 Quantitative variation in nuclear DNA amount in a number of major groupings of plant and animal species. [From Sparrow, A. H., Price, H. J. and Underbrink, A. G. (1972) *Brookhaven Symp. Biol.*, **23**, 451–94]

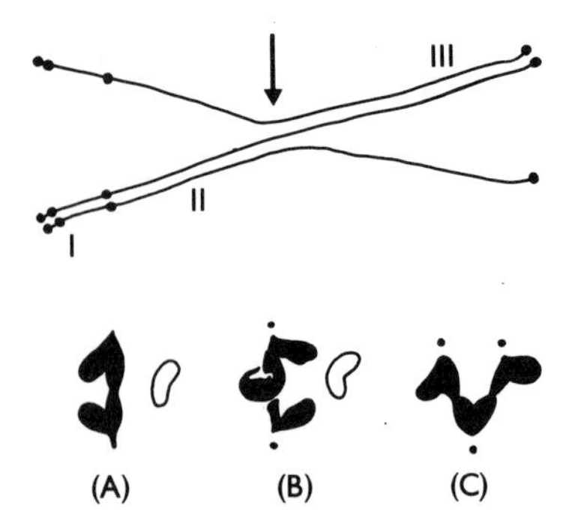

Fig. 4.3 Pachytene pairing in a triploid. Positions of centromeres are in the vicinity of the arrow. One terminal chiasma only, e.g. in region I, gives a rod bivalent and a univalent at first metaphase (A); two chiasmata, one terminal and one interstitial, e.g. at I and II, a rod bivalent with two chiasmata in one arm (B). Two chiasmata at I and III produce a trivalent (C).

population from which they derive. From the cytological standpoint the extra sets of chromosomes pose no problems at mitosis. The orientation of the chromosomes on the spindle at metaphase and their separation at anaphase are effective and efficient. At meiosis, however, the pachytene associations involve three or more homologous chromosomes. Following chiasma formation they may be transformed to multivalent rather than bivalent configurations. The formation of multivalents, in turn, affects the segregation at first anaphase of meiosis and, as a result, the distribution of genetic material to the gametes.

(i) *Meiosis in triploids.* Although the three homologues of each member are associated during pachytene the association at any one chromosome segment is restricted to a pair (Fig. 4.3). Recent electron microscope investigations, it is true, have revealed intimate contact between all three homologues (Moens, 1969b) but such contacts are restricted to very short segments and contact is between pairs for the greater part of the chromosomes. The metaphase configuration resulting from this association depends on the number and the position of chiasmata within it. Evidently with no chiasmata the result will be three univalents, with one chiasma a rod bivalent. With two chiasmata involving the same pair of chromosomes we get a bivalent and a univalent. With two chiasmata involving all three of the chromosomes the result is a trivalent. Loss of univalents during meiosis will, of course, be a cause of inviability among gametes. Even with complete trivalent formation, however, their asymmetry at metaphase results in the production of gametes with varying chromosome numbers. Fig. 4.4 shows all four types of arrangement of three trivalents in a triploid $2n=3x=9$. Twenty five per cent of the gametes contain either haploid or diploid (euploid) chromosome complements. The remainder are aneuploid with either one or two of the

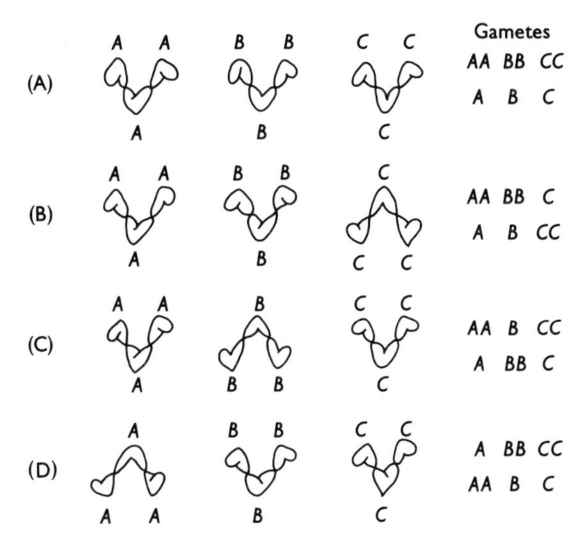

Fig. 4.4 The four alternative types of orientation of trivalents at first metaphase of meiosis in a triploid $2n=3x=9$. Only one (A) produces euploid gametes. The gametes from the remainder are aneuploid which, as a result, would normally be inviable or ineffective.

chromosomes in double dose. This means that the genes from the different chromosomes are unequally represented, resulting in a genetic imbalance that is often lethal (see Darlington and Mather, 1949). Assuming, in this hypothetical case, all the aneuploid gametes to be lethal only 25% of the gametes would be viable. In fact an analysis in a $3x=9$ triploid *Crepis capillaris* by Chuksanova (1939) showed precisely this degree of gamete viability.

Basic number. With three trivalents per cell we have only 25% euploid, viable gametes, corresponding with the percentage of cells in which all three trivalents are orientated in the same way relative to the spindle poles. In a triploid with a basic number of 4, i.e. $3x=12$ the probability of such orientation is clearly less (12.5%): and less still with higher basic numbers. Indeed triploids with basic numbers of 6 or more are almost completely sterile. Some varieties of apples, such as Blenheim Orange and Bramley's Seedling are triploids ($3x=51$). The seed set is less than 5% and the percentage of germinating seeds even lower (Crane and Lawrence, 1934).

In almost all odd numbered polyploids, $5x$, $7x$ etc., the degree of sterility is extremely high. As in the triploid described the main cause is the production of aneuploid, genetically imbalanced gametes. There are rare exceptions. In *Hyacinthus*, for example, aneuploid gametes are relatively viable and fertile (Darlington, Hair and Hurcombe, 1951).

For this reason the species is represented by an unusually high number of aneuploid individuals.

(ii) *Meiosis in tetraploids*. Since each chromosome is represented by four homologues the pachytene associations are made up of all four, but again the contact at any one segment is restricted to a pair. As with the triploid the frequency of different kinds of configurations at metaphase: univalents, bivalents, trivalents and quadrivalents, will depend on the frequency and location of chiasmata within the pachytene associations (Fig. 4.5). Loss of chromosomes in the form of univalents, including those accompanying trivalents, are a cause of aneuploidy and consequently of inviable gametes. A high frequency of trivalents could also cause aneuploidy, as in triploids, on grounds of asymmetry of the metaphase configurations and unequal anaphase separation. In many species, however, trivalents are conspicuously rare in autotetraploids. This is especially true for species with chiasmata located mainly in distal segments of chromosomes. With the formation of 3 or 4 chiasmata near the ends of the chromosomes in pachytene associations of four, only quadrivalents and bivalents are possible. Another reason for the absence of trivalents is that the four homologous chromosomes may associate as two pairs. In most pollen mother cells of autotetraploid rye, for example, the four chromosomes of one of the seven homologous sets associate exclusively in pairs at pachytene (Timmis and Rees, 1971). Where the chiasma frequency is reasonably high (of the order of three or more per association of four) trivalents and univalents will be very rare. Equally important the quadrivalents formed, as well as the bivalents, separate in a perfectly regular, 2×2 fashion at anaphase to produce euploid, viable gametes.

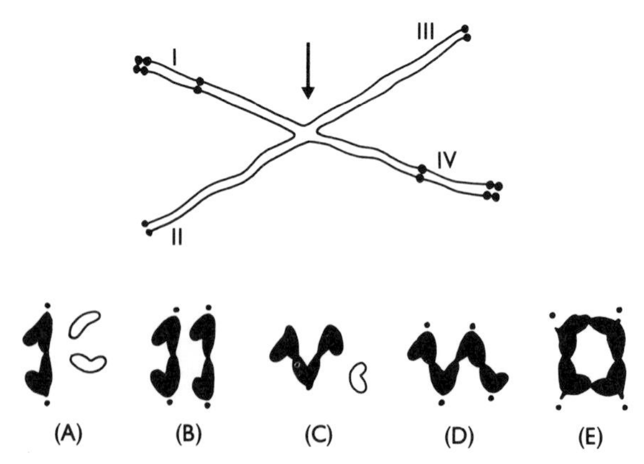

Fig. 4.5 Pachytene pairing in a tetraploid. The arrow indicates the position of the centromeres. The first metaphase configurations below result from the formation of: (A) one terminal chiasma, e.g. at I; (B) two chiasmata, one at I and IV (or at II and III); (C) two chiasmata at I and II or III (or IV and II or III); (D) three terminal chiasmata (e.g. at I, II and III) and (E) four terminal chiasmata. Univalents unshaded.

For this reason autotetraploids are far less infertile than triploids and, to generalize, even numbered autopolyploids are more fertile than odd numbered, although never as fertile as diploids.

Chromosome pairing and fertility. Univalents and trivalents are the chief causes of aneuploidy and of inviability among gametes produced by autotetraploids. It follows that a reduction in the frequency of univalents and trivalents should improve fertility. Whether this be achieved at the expense of increasing quadrivalents or bivalents should be immaterial because both disjoin regularly at first anaphase, at least in species with distally localized chiasmata. We have indicated that the frequency of trivalents and univalents is high in organisms with low chiasma frequencies. In such cases we should expect an improvement in fertility, therefore, by increasing the chiasma frequency. Investigations in rye and in *Lolium* (Hazarika and Rees, 1967; Crowley and Rees, 1968) confirm that an increase in the chiasma frequency is indeed accompanied as expected by a reduction in univalents and trivalents and also by an increase in quadrivalents. They also show an increase in fertility as measured by the seed setting.

Table 4.1 The frequencies of quadrivalents (IVs), trivalents (IIIs), bivalents (IIs) and univalents (Is) in pollen mother cells of tetraploid populations of *Lolium perenne* ($2n = 4x = 28$) and *Secale cereale* ($2n = 4x = 28$) before and after selection for improvement in fertility. Selection in *Lolium* was over five generations and over 19 generations in *Secale*. Note the increase in the frequencies of quadrivalents and the decrease in frequencies of trivalents and univalents among selected populations. [*Lolium* data from Crowley, J. G. and Rees, H. (1968) *Chromosoma*, **24**, 300–8; *Secale* data from Dr. Gul Hussain, by kind permission]

	Unselected				Selected			
	IV	III	II	I	IV	III	II	I
Lolium perenne	3.01	0.32	7.24	0.49	4.07	0.10	5.67	0.10
Secale cereale	2.97	0.68	9.04	0.75	5.12	0.34	7.74	0.19

Investigations of autotetraploid rye and *Lolium* varieties selected for improvement in fertility provide further confirmation that reducing the number of trivalents and univalents and increasing the quadrivalent frequency by increasing the chiasma frequency is effective (Table 4.1). The quadrivalents are more numerous as a direct result of an increase in chiasma frequency. From the standpoint of fertility an increase in bivalent frequency would have been equally acceptable. Even so it is worth emphasizing that multivalents, in themselves, provided they disjoin with regularity as do the quadrivalents in rye and *Lolium* are no barrier to fertility.

Finally, it is important to appreciate that infertility in polyploids is not solely due to the production of aneuploid gametes. Different tetraploid strains of rye vary considerably in seed set despite the fact that they display similar patterns of chromosome pairing at first anaphase of meiosis and each produces much the same proportion of aneuploid gametes. The variation is evidently due to *genic* as distinct from *chromosomal* causes (Hazarika and Rees, 1967).

(b) *Allopolyploids*

At meiosis in hybrids between diploid species the pachytene pairing is often ineffective with the result that chiasma formation is severely

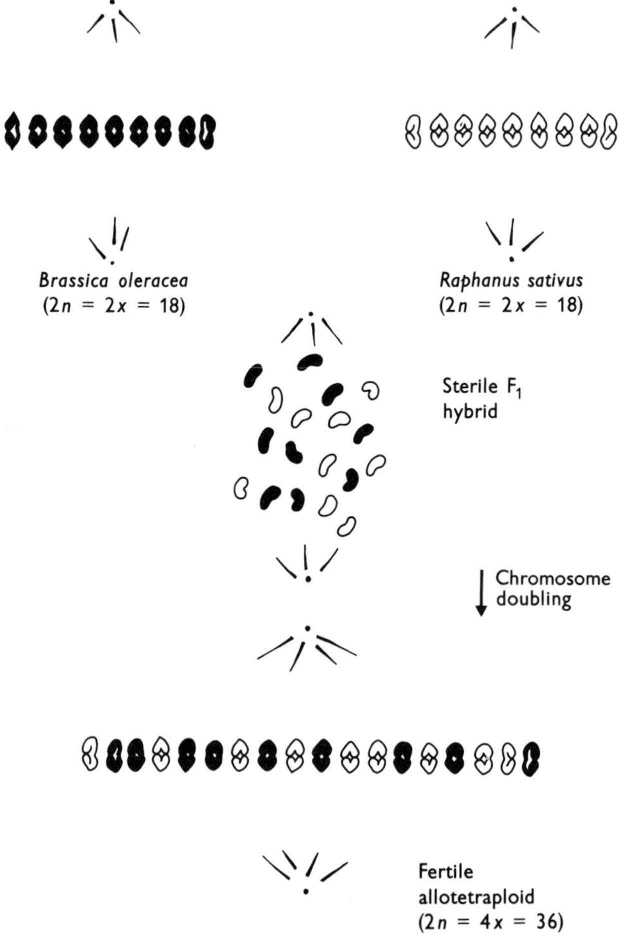

Fig. 4.6 Chromosome pairing at first metaphase of meiosis in *Brassica oleracea*, *Raphanus sativus*, their F_1 diploid hybrid and in the allotetraploid. The lack of homology between *B. oleracea* and *R. sativus* chromosomes is the cause of failure of pairing in the diploid hybrid and of bivalent formation in the allotetraploid.

Table 4.2 Polyploidy in flowering plants

Species (N = natural, C = cultivated)	Common name	Constitution	Origin
Allopolyploids			
Galeopsis tetrahit (N)	hemp nettle	$2n=4x=32$	*G. pubescens* × *G. speciosa* ($2x=16$) ($2x=16$)
Raphano-Brassica (C)		$2n=4x=36$	*Raphanus sativus* × *Brassica oleracea* ($2x=18$) ($2x=18$)
Brassica napus (N)	rape	$2n=4x=38$	*B. oleracea* × *B. campestris* ($2x=18$) ($2x=20$)
Triticale (C)		$2n=8x=56$	*Triticum aestivum* × *Secale cereale* ($6x=42$) ($2x=14$)
Triticum aestivum (C)	wheat	$2n=6x=42$	*T. monococcum* × *T. speltoides* × *Aegilops squarrosa* ($2x=14$) ($2x=14$) ($2x=14$)
Primula kewensis (C)		$2n=4x=36$	*P. floribunda* × *P. verticillata* ($2x=18$) ($2x=18$)
Nicotiana tobacum (C)	tobacco	$2n=4x=48$	*N. sylvestris* × *N. tomentosiformis* ($2x=24$) ($2x=24$)
Autopolyploids			Method of propagation
Beta vulgaris (C)	sugar beet	$2n=3x=27$	seed, by crossing diploid × tetraploid
Malus pumila (C)	apple	$2n=3x=51$	grafting
Secale cereale (C)	rye	$2n=4x=28$	seed
Allium tuberosum (N)		$2n=4x=32$	bulbs
Ranunculus ficaria (N)	celandine	$2n=2x, 4x, 5x, 6x$ $=24, 32, 40, 48$	bulbils
Solanum tuberosum (C)	potato	$2n=4x=48$	tubers
Hyacinthus orientalis (C)	hyacinth	$2n=3x, 4x=24, 32$	bulbs
Fritillaria camschatensis (N)		$2n=3x=36$	bulbs

reduced and chromosomes are found unpaired, as univalents, at first metaphase. Such hybrids whether they be mules or F_1s from crossing *Brassica* species are virtually sterile. It was in the genus *Brassica* that Karpechenko (1928) first explained how the doubling of the chromosome number in a hybrid produced a fertile tetraploid (Fig. 4.6). The chromosomes of the intergeneric hybrid between *Raphanus sativus* (radish) × *Brassica oleracea* (cabbage) show very little pairing at meiosis in the hybrid between them. They are not homologous, by which is meant that they are not sufficiently alike structurally to achieve effective pairing. This is not to say, of course, that they are structurally or genetically unrelated and for this reason the members of each 'pair' are properly described as *homoeologous* as distinct from homologous. As Fig. 4.6 shows, doubling the chromosome number permits pairing between strictly homologous chromosomes and since pairing is restricted to homologues only bivalents appear at first metaphase. In the absence of multivalents separation at first anaphase of meiosis is regular, the gametes are euploid (diploid) and viable. Such a tetraploid is an allotetraploid or amphidiploid. Other examples of allopolyploids are given in Table 4.2. Their fertility is high and comparable to that of diploids.

(c) *Segregation in polyploids*

Odd numbered polyploids are, almost invariably, highly infertile. Their propagation and survival under natural or cultivated conditions is consequently dependent upon asexual means. Examples are given in Table 4.2. Such clonal propagation means that progenies are genetically identical with one another and with their parents. In autotetraploids the segregation of the four alleles of a gene to give diploid gametes is random. The gametes formed from the alleles a^1, a^2, a^3, a^4 will be of the constitution a^1a^2, a^1a^3, a^1a^4, a^3a^4, a^2a^4, a^2a^3 and will occur with equal frequency. Gametes produced by quadruplex (*AAAA*), triplex (*AAAa*), duplex (*AAaa*), simplex (*Aaaa*) and nulliplex (*aaaa*) tetraploids are given in Table 4.3. There is one qualification to be made. Triplex and simplex tetraploids in the Table give no double recessives (*aa*) or dominants (*AA*) respectively. Yet such gametes do arise (Mather, 1936). For example, *aa* gametes are produced by a triplex when: (1) a chiasma forms between the centromere and the *A* locus; and (2) the orientation of the chromosomes at first and second anaphase is such that *a* and *a* move to the same

Table 4.3 Segregation in autopolyploids

		Gametes	Phenotypic ratio
Quadruplex	*AAAA*	*AA*	1
Triplex	*AAAa*	1*AA*:1*Aa*	1:1
Duplex	*AAaa*	1*AA*:4*Aa*:1*aa*	5:1
Simplex	*Aaaa*	1*Aa*:1*aa*	1:1
Nulliplex	*aaaa*	*aa*	1

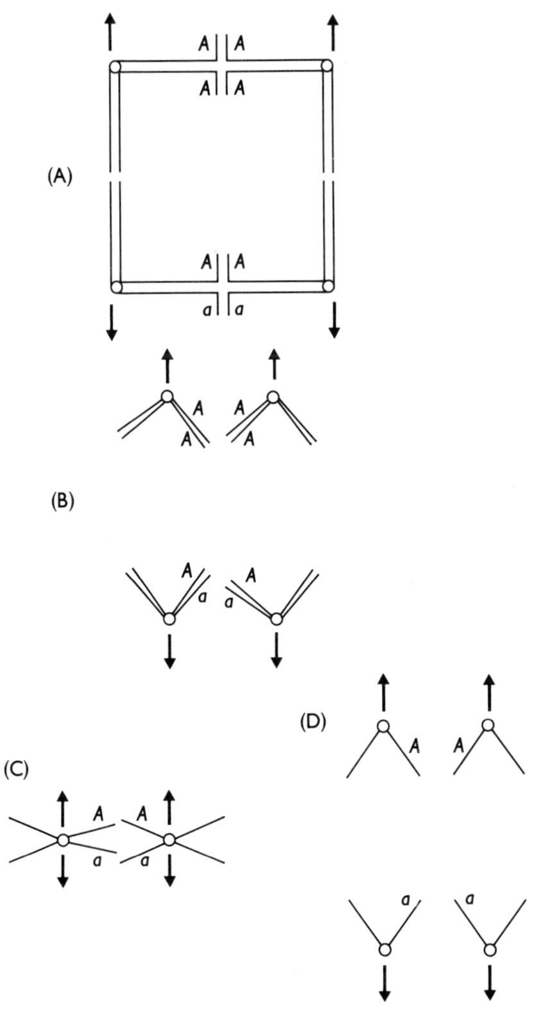

Fig. 4.7 Chromatid segregation in an autotetraploid. In a triplex *AAAa*, a chiasma between the centromere and the locus *A* (A) is followed by reductional separation at first anaphase (B). Orientation at second metaphase (C) is such that *aa* are drawn to the same pole at the second anaphase (D). The frequency with which a chiasma forms between the centromere and the locus will depend upon the distance between them. If a chiasma does form, separation at first anaphase will be reductional in half the cases and orientation at the second metaphase will, again in half the cases, give *aa* gametes. While the kinds of separation at first and second anaphase are predictable in any situation, the frequency with which the chiasma forms between the centromere and the locus can only be established by experiment. Therefore the amount of chromatid interference varies from one locus to another.

poles (Fig. 4.7). The frequency with which such gametes are produced will depend on the frequency with which chiasmata occur between *A* and the centromere and hence upon the distance between the centromere and the locus.

In allotetraploids, derived from chromosome doubling of a diploid hybrid, the gametes produced will be of uniform constitution. For the alleles a^1a^1 on a pair of chromosomes from the one parent and a^2a^2 on the homoeologous pair from the other, the gametes will be invariably a^1a^2. Progenies of the allotetraploid will therefore be similar to one another and to the parents. In other words the species hybrid derived by allopolyploidy is immediately true breeding.

While natural allopolyploids arise by chromosome doubling in diploid hybrids it is possible to synthesize allopolyploids by crossing autotetraploids from different species. In this case homologous chromosomes may be heterozygous, e.g. a^1a^2, a^1a^2 and such an allopolyploid would not, of course, breed true on selfing.

(d) *The classification of polyploids*

We have treated autopolyploids and allopolyploids, in terms of chromosome pairing and segregation, as sharply distinct and disparate forms. The distinction is rarely, if ever, as sharp, and homoeologous chromosomes are not always sufficiently different in structure to prevent pairing between them, so that multivalents as well as bivalents may well be formed at meiosis in polyploids produced from species hybrids. *Triticum aestivum*, the cultivated bread wheat is a hexaploid ($6x=42$) made up from three diploid genomes, *AA*, *BB* and *DD* which are derived from the primitive diploid wheats *Triticum monococcum*, a species related to *Aegilops speltoides* and *Triticum squarrosa* (Fig. 4.8). At metaphase of meiosis, only bivalents are formed and separation is regular to give viable gametes, each within an *A*, *B* and *D* genome making up 21 chromosomes. On the face of it this is a classical example of chromosome pairing at meiosis in an allopolyploid. Riley and Chapman (1958), however, have shown that the restriction of pairing to strict homologues is not entirely accounted for by structural differences between homoeologous chromosomes. A gene, or cluster of genes, in the long arm of chromosome 5 of the *B* genome is an essential component of the mechanism ensuring pairing between homologues only. In the absence of this gene or gene cluster (as in a wheat plant carrying only telocentrics for the short arm of the 5*B*-chromosome) there is pairing between homoeologues as well as homologues. In short there are multivalents at metaphase of meiosis as well as bivalents. It is probable that such diploidizing genes or gene complexes operate in other allopolyploids reinforcing the structural differences between chromosomes to prevent pairing and chiasma formation between other than strictly homologous chromosomes. This implies that non-homology is not, in itself, sufficient to ensure only bivalent formation at meiosis. It is clear as well that the degree of homology between chromo-

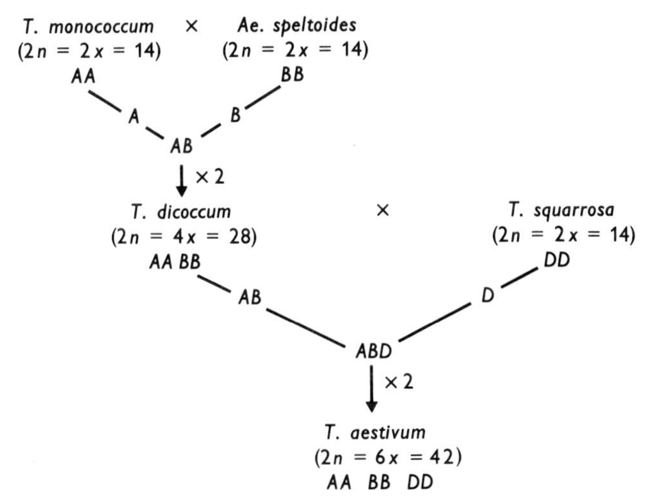

Fig. 4.8 The origin of the allohexaploid cultivated wheat, *Triticum aestivum*, by hybridization and chromosome doubling in the ancestral species and hybrids. We have given *Ae. speltoides* as the source of the *B* genome. An *Ae. speltoides*-like species would be more accurate.

somes from different species will vary from slight to substantial. Consequently, at meiosis in even numbered polyploids from hybrids between species, we may find a range of events from widespread multivalent formation typical of autopolyploids to bivalent formation only, typical of allopolyploids.

(e) *The polyploid phenotype*

In plants, polyploids, in comparison with their diploid ancestors, are generally slower in initial growth and later in maturing, more tolerant of extremes of temperature and rainfall and have larger and thicker leaves. As Stebbins (1950) makes clear there are abundant exceptions to this broad generalization. In *Tradescantia*, for example, diploids and tetraploids are indistinguishable on the basis of external appearance (Swanson, 1960). The same is true of diploid and polyploid forms of the wild potato, *Solanum tuberosum* (Dodds, 1965). The effects of doubling the chromosome number may also vary from one genotype to the other within a population. In rye, for example, the dry weights of tetraploids derived from some genotypes are twice that of the diploid, in others the dry weights of diploids and tetraploids are similar (Hutto, 1974). The generalization about the effects of polyploidy upon growth and development is, as we have stressed, subject to many qualifications. We shall return to this issue later (p. 96).

(f) *Variability*

Lewis (1967) has pointed out that a substantial portion of both root and shoot tissue in plants is composed of polyploid cells. He argues that any physiological advantages conferred by polyploidy upon individuals are readily achieved by the induction and restriction of polyploidy to differentiated, non-dividing cells. Above all this would avoid the complications at meiosis which cause infertility in autopolyploids. On this basis he considered that from the standpoint of adaptation in evolution the importance of polyploidy rests mainly on genetical rather than physiological criteria. He argues that polyploidy, above all, serves to maintain a high level of heterozygosity following hybridization. For a single gene with two alleles, *A* and *a*, the relative frequencies of homozygotes and heterozygotes in random mating populations of diploids, auto- and allotetraploids are given in Table 4.4. Given heterozygous advantage in the form of heterosis the generation and maintenance of heterozygosity in polyploids may well be of paramount importance. It will be observed that the combination *Aa* in the allotetraploid, following hybridization between species carrying *AA* and *aa* respectively, is permanently fixed and even with inbreeding the heterozygosity would be retained and maintained by the restriction of pairing to strictly homologous chromosomes, in this case carrying, on the one hand, *AA* and on the other, *aa* (see also Rees and Jones, 1971).

(g) *Polyploidy in animals*

In our brief account of polyploidy and of polyploids we have confined our observations to plants. There are two good reasons for this. The first is that, in higher organisms especially, polyploidy is rare in animal species and, consequently, most of the investigations and most of the information on polyploidy refers to plants. The two reasons given to

Table 4.4 The relative frequencies of heterozygotes and homozygotes in populations of diploids, autotetraploids and allotetraploids: assuming a 1:1 distribution of alleles (*A* and *a*), random mating and no selection. [From Rees, H. and Jones, R. N. (1971) *Cellular Organelles and Membranes in Mental Retardation*, Churchill Livingstone]

	Allelic constitution	Heterozygotes
Diploid	*AA* 1 *Aa* 2 *aa* 1	50%
Autotetraploid	*AAAA* 1 *AAAa* 4 *AAaa* 6 *Aaaa* 4 *aaaa* 1	87.5%
Allotetraploid	*AAaa* 1	100%

account for the rarity of polyploidy among animals are: (1) that individuals are more often unisexual rather than hermaphrodite as are members of most species of higher plants; and (2), that polyploidy may upset sex determination, in particular where this depends on a balance between *X*-chromosomes and autosomes as in *Drosophila*. The unisexuality of individuals is a restriction on the establishment of polyploidy in the following way. When a tetraploid arises by chromosome doubling, be it male or female, its diploid gametes can fertilize or be fertilized only by haploid gametes. The offspring will be triploid and sterile, so that in the absence of asexual means of propagation the polyploidy cannot be maintained. In hermaphrodite species of animals, among earthworms and snails for example, there is a possibility, as in plants, of self fertilization and thereby the establishment of fertile polyploid forms. As for sex determination Bridges (1922) was among the first to establish in *Drosophila*, that an *X*/autosome ratio of 0.5 (*X*/*AA* in the normal diploid) irrespective of the presence or absence of *Y*-chromosomes, determined maleness; and an *X*/autosome ratio of 1.0, (*XX*/*AA* in the normal diploid) determined femaleness. Intermediate ratios resulted in the development of intersexes. On this basis tetraploids derived from chromosome doubling to give *XX*/*AAAA* and *XXXX*/*AAAA* from diploid males and females respectively would develop as normal males ($X/A = 0.5$) and females ($X/A = 1.0$). With regular segregation to give *X*/*AA* and *XX*/*AA* sperm and eggs, however, fertilization would result in *XXX*/*AAAA* ($X/A = 0.75$) sterile intersexes. In other groups, however, the determination of sex should not be upset by polyploidy. In the mammals, for example, the sex determining mechanism is not based on an *X*/autosome balance. In the main it depends on the presence of *Y* to give males, on the absence of a *Y* to give females. Thus an *XY* mouse is a male, an *XO* mouse (missing a *Y*-chromosome) is a female and a fertile female at that (Morris, 1968). The *X*/*A* ratio is, of course, the same in both. Yet polyploid adult individuals, let alone populations, are virtually unknown in mammals. It is clear, therefore, that factors other than those we have considered above restrict the establishment of polyploidy in mammals and other groups. In humans for example only a very small number of polyploid fetuses survive anywhere near a full term pregnancy whereas 0.6–0.9% of fertilizations give rise to polyploid embryos (Hamerton, 1971). Of four triploid births recorded two were stillborn and the other two died soon after birth (Schindler and Mikamo, 1970). Evidently there is a physiological restriction upon the growth and the development of polyploids among higher animals. The causes are not known. Incompatibility between the polyploid embryo and the diploid mother may be one.

(c) *Aneuploidy*

Aneuploids, like polyploids, are again commoner among higher plants than higher animals. The higher incidence of polyploid higher plants is one cause. Most aneuploids are the direct result of the irregular disjunction of chromosomes at meiosis in polyploids which give rise to aneuploid gametes. Among the progenies of autotetraploid *Lolium perenne* from 6 to 23% are aneuploids (Ahloowalia, 1971). Even under experimental conditions, however, only a limited fraction of aneuploids are effective as gametes or viable as zygotes. Of these the 'fittest' have chromosome numbers close to the euploid values, e.g. $4x+1$ and $4x-1$, where the genetic imbalance is minimal. For the same reason the addition of one chromosome to the diploid complement, to give $2x+1$ trisomics, is likely to cause less imbalance than the addition of two or three chromosomes. This is not to say, of course, that trisomics are free of the symptoms of imbalance. On the contrary they are almost invariably abnormal in development and frequently highly infertile. As one might expect the imbalance generated by trisomy is partly dependent on the particular chromosome involved. This is reflected by surveys within the cultivated tomato (Table 4.5). The table shows that the frequency of different kinds of trisomics among the progenies of triploids varies considerably, indicating that the viability of trisomics, as gametes or zygotes, is closely related to the particular chromosome addition. Trisomics for chromosomes 4 and 5 are relatively numerous in comparison with trisomics for 1 and 11. This fact also makes clear that the degree of imbalance is not related directly to chromosome length (see Table 4.5) and therefore the amount of genetic material which it carries. This is just as true for humans as for tomatoes. Among the most common aneuploids of man are those involving the *X*-chromosome which is by no means the smallest of the complement.

Table 4.5 The relative frequencies of the 12 trisomics in the progenies of triploid tomatoes. [From Rick, C. M. and Barton, D. W. (1954) *Genetics*, **39**, 640–66]

Chromosome number	Frequency of trisomics as percentage of total progeny	Chromosome length in microns
1	0.5	52.0
2	3.3	42.1
3	1.6	40.3
4	9.9	35.0
5	6.3	31.4
6	0.7	31.3
7	3.5	27.5
8	4.6	27.5
9	2.7	26.8
10	5.5	25.7
11	0.1	24.8
12	4.0	22.5

A few among higher plant species are exceptional in displaying tolerance to aneuploidy. The best studied is the common hyacinth, *Hyacinthus orientalis*. Garden varieties include diploids ($2x = 16$), triploids ($3x = 24$) and tetraploids ($4x = 32$). In addition, however, there is a whole range of aneuploids with chromosome numbers of 17, 19, 20, 21, 23 etc. Even here, however, some aneuploid combinations are more common, more tolerated than others. One supposes that in *Hyacinthus* efficient gene action is less dependent upon the interaction of genes on *different* chromosomes than is the case in other species (Darlington, Hair and Hurcombe, 1951). One must emphasize, however, that this situation in hyacinth is quite uncharacteristic. In the vast majority of species the consequences of aneuploidy are so deleterious on grounds of imbalance as to render aneuploidy of little consequence from the adaptive or evolutionary standpoint.

_B_-chromosomes

The chromosome complements of individuals within a species exhibit a remarkable degree of constancy in terms both of the number and form of their chromosomes. A notable exception to this general rule is provided by *B*-chromosomes. These chromosomes are supernumerary to the normal basic complement and are found in numerous species of both plants and animals. They are calls *B*s to distinguish them from the normal (*A*) chromosomes and they are characteristically smaller than the *A*-chromosomes, not homologous with any of the *A*s in that they never pair with *A*s at meiosis, variable in number both between populations within species and among individuals within populations. The *B*-chromosomes, on grounds of their size and pairing behaviour at meiosis, have diverged substantially in structure from the *A*-chromosomes from which presumably they derive. The DNA which they contribute to the nucleus, therefore, does not represent a precise copy of the nuclear DNA fraction as is the case with aneuploidy. Their evolution involves a qualitative change superimposed upon a change in amount. Even in those populations with a high frequency of *B*s, however, there are some individuals without *B*s. Evidently they are dispensable, and not essential for normal growth, development and reproduction. For this reason it is not surprising that they were often considered to be genetically inert and of little adaptive or functional significance. This view was reinforced by lack of evidence of any major gene effects attributable to *B*s. More recently, however, it has become increasingly clear that, although dispensable, the effects of *B*s upon the phenotype are manifold and in some instances quite startling. The indications are that they are of substantial adaptive significance.

(a) *Occurrence*

B-chromosomes were first distinguished and classified by Longley (1927) and Randolph (1928) in maize. The numbers of higher plants and

Table 4.6 *B*-chromosome distribution among groups of plant and animal species

Group	No. of species
PLANTS	
Gymnosperms	7
Angiosperms	
Dicotyledons	318
Monocotyledons	319
Total	644
ANIMALS	
Platyhelminths	2
Molluscs	2
Insects	158
Amphibians	4
Reptiles	1
Mammals	11
Total	178

of animal species now known to carry *B*-chromosomes is considerable, as Table 4.6 shows. Contrary to previous assertions, *B*-chromosomes are found in about the same proportion of polyploid as of diploid species. In fact in some cases, such as in *Leucanthemum* populations in Yugoslavia, *B*s are found only within the polyploid populations. In other species, e.g. *Ranunculus ficaria* in Britain, the opposite situation prevails and the *B*s are restricted to diploids. *B*-chromosomes have not been reported in inbreeding species.

Table 4.7 gives the range of variation in numbers of *B*s found among natural populations. Under experimental conditions even higher numbers, up to 34 in *Zea mays* and 22 in *Centaurea scabiosa*, are reported. These two examples serve to illustrate the very considerable amount of additional DNA that may be accumulated within the nucleus due to *B*-chromosomes. In maize the addition of 34 *B*s to the diploid complement of 20 *A*-chromosomes is equivalent to the amount of DNA that would be added by having the *A*-chromosome set multiplied five times, equivalent that is to pentaploidy.

(b) *Inheritance*

Somatic cell division. In many species the transmission of *B*-chromosomes during somatic cell division is normal and in consequence their frequency is constant throughout the somatic cells within the organism. Among plants this is true, for example, in *Secale cereale*, *Anthoxanthum odoratum*, and *Clarkia elegans*; among animals, for *Acrida lata*, *Myrmeleotettix maculatus*

Table 4.7 The range of *B*-chromosome numbers in some plant and animal species

Species	Number of *B*-chromosomes																	
	0	1	2	3	4	5	6	7	8	9	10	11	12	13	14	15	16	Σ
Centaurea scabiosa																		
Scandinavia and Finland	5688	611	564	333	218	128	63	64	39	8	15	1	1	1	2	1	1	7738
Festuca pratensis																		
Sweden	854	97	68	8	11	3	3	2	–	–	–	–	–	–	–	–	–	1046
Crepis conyzaefolia																		
Italy	133	31	11	6	1	–	–	–	–	–	–	–	–	–	–	–	–	182
Tainia laxiflora																		
Japan	5	10	49	17	24	5	2	1	4	2	–	–	–	–	–	–	–	119
Ranunculus ficaria																		
Britain	45	8	12	15	5	2	9	3	–	–	–	–	–	–	–	–	–	99
Reithrodontomys megalotis																		
U.S.A.	39	89	94	45	15	2	0	1	–	–	–	–	–	–	–	–	–	285
Myrmeleotettix maculatus																		
Wales	90	41	9	–	–	–	–	–	–	–	–	–	–	–	–	–	–	140

and *Pseudococcus obscurus*. In other cases the *B*s are unstable at mitosis so that their frequency varies between different tissues and organs. The instability takes the form of lagging at anaphase or of non-disjunction. At the time of flower initiation in *Crepis capillaris* the *B*-chromosomes regularly undergo non-disjunction in shoot meristems. The non-disjunction is polarized such that the *B*-chromosome frequency is increased in the germ line cells. In *Xanthisma texanum*, *Aegilops speltoides* and *Haplopappus gracilis*, *B*-chromosomes are excluded from the roots but are present in the shoots and inflorescences. Among animals species mitotic instability is common during the development of the testis, leading to variation in *B* frequency among cells within and between the follicles, e.g. *Locusta migratoria*, *Pantanga japonica* (Kayano, 1971; Sannomiya, 1962).

Germ line cells. When a rye plant carrying one *B*-chromosome is crossed to a rye plant without *B*s, the progeny resulting from this cross consist mainly of 0*B* and 2*B* individuals. If two rye plants each carrying 2*B* chromosomes are crossed together the bulk of the offspring each possess 4*B*s. Clearly the mode of inheritance of these additional *B*-chromosomes is non-Mendelian and leads to a numerical accumulation of the *B*s over a succession of generations with a preponderance of the even numbered combinations. The mechanism which accounts for this unorthodox mode of inheritance is now firmly established for rye. Meiosis itself is fairly regular (see Fig. 4.23). With 1*B*, segregation in pollen mother cell is such that two of the four products carry 1*B* and the other two none. If 2*B*s are present they form a bivalent at the first division and segregate one *B* to each of the four microspores of the tetrad. There is never pairing between *A*s and *B*s. It is during the first mitotic division of the microspore, i.e. first

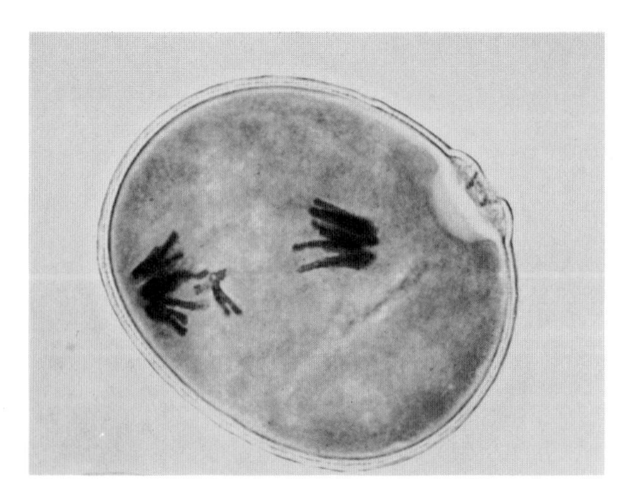

Fig. 4.9 Anaphase of the first pollen grain mitosis in rye showing a single *B*-chromosome undergoing non-disjunction.

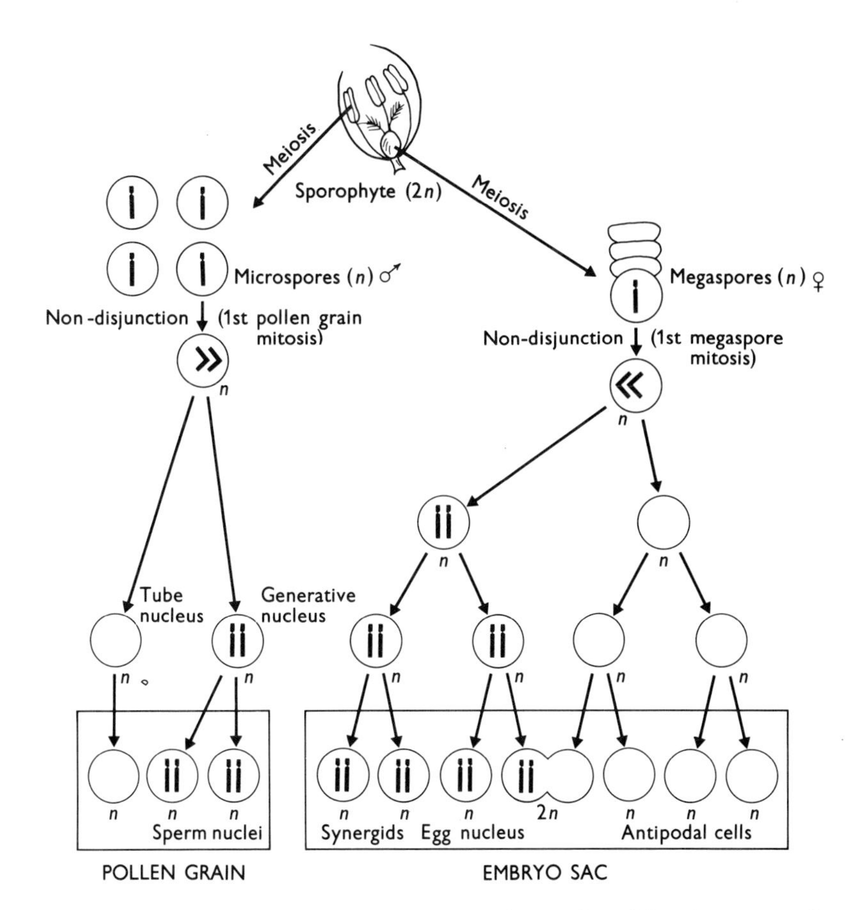

Fig. 4.10 The transmission of *B*-chromosomes into male and female gametes of a rye plant with two *B*-chromosomes. The non-disjunction at the first pollen grain and at the first megaspore cell mitosis foster the accumulation of *B*s among progenies. Note that sperm and egg nuclei both contain two *B*-chromosomes; hence a cross between rye parents each with two *B*-chromosomes yields progeny with 4 *B*-chromosomes. Only the *B*-chromosomes are drawn.

pollen grain mitosis, that the *B*s behave abnormally. The haploid complement of *A*-chromosomes divides mitotically with one set passing into the generative nucleus and the other to the tube nucleus of the developing male gametophyte. At the anaphase of this first pollen grain mitosis, however, the daughter chromatids of the *B*-chromosomes fail to disjoin even though the centromeres have divided, and they pass together preferentially into the generative nucleus (Fig. 4.9). They divide quite normally at the second pollen grain mitosis. In the embryo sac a similar sequence of events takes place with the *B*s undergoing a polarized non-disjunction at the first mitosis of the functional megaspore, again in such a direction that they ensure their perpetuation in the germ line. The scheme is outlined in Fig. 4.10.

Some element of non-disjunction and accumulation is found in most of the plant species investigated, although the details of the mechanism vary somewhat from one species to another. In *Zea mays*, for example, polarized non-disjunction occurs at the *second* pollen grain mitosis, while transmission through the female gametophyte is quite normal. Preferential fertilization by the *B*-carrying male gametes, is a further cause of distortion from the normal Mendelian pattern in maize (Roman, 1948).

In rye, non-disjunction is brought about by failure of chromatid separation at two sites, one each side of the centromere. While the 'sticking' of chromatids is near the centromeres, the capacity for 'sticking'

Table 4.8 Mechanisms of *B*-chromosome accumulation in plants

Mechanism	Species
Preferential meiotic segregation in egg mother cells	*Lilium callosum* *Tradescantia virginiana* *Phleum nodosum*
Preferential fertilization by *B*-carrying male gametes	*Zea mays*
Directed non-disjunction:	
First pollen grain mitosis	*Aegilops speltoides* *Alopecurus pratensis* *Anthoxanthum aristatum* *Briza media* *Dactylis glomerata* *Deschampia caespitosa* *Festuca pratensis* *Haplopappus gracilis* *Phleum phleoides*
Second pollen grain mitosis	*Zea mays*
First pollen grain mitosis and first egg cell mitosis	*Secale cereale*
First pollen grain mitosis of extra divisions	*Sorghum purpureo-sericeum*
Somatic non-disjunction coincident with flower initiation	*Crepis capillaris*
No apparent mechanism	*Centaurea scabiosa* *Poa alpina* *Ranunculus ficaria* *Xanthisma texanum*

is under the control of a 'gene' located in the terminal block of heterochromatin in the long arm. Following deletion of the controlling site the *B*-chromosome loses its capability of undergoing non-disjunction. Ward (1973) has established that the controlling site for non-disjunction in maize lies in a short euchromatic segment distal to the major heterochromatic region of the long arm, at which the 'sticking' takes place. In maize the *B*-chromosomes are also the cause of sticking at heterochromatic 'knobs' in the *A*-chromosomes at the second anaphase of meiosis with the result that segments of *A*-chromosomes carrying knobs are eliminated (Rhoades and Dempsey, 1973).

The accumulation mechanism in *Lilium callosum* differs from those in rye and maize in that it operates at meiosis and not in the post-meiotic divisions in the gametophyte. Transmission through the pollen is normal, but at female meiosis there is a preferential distribution of *unpaired Bs* in egg mother cells towards that half of the spindle nearest the micropylar end of the embryo sac. In *Lilium* the functional megaspore is proximal to the micropyle so that the *B*s are accumulated in the egg cells. A summary of some of the systems of *B*-chromosome inheritance in plants is given in Table 4.8. By far the commonest is that involving polarized non-disjunction at the first pollen grain mitosis with normal transmission through the female.

B-chromosomes of course cannot accumulate indefinitely. The tendency to increase in number over successive generations is countered, as we shall see, by a severe impairment of reproductive fitness when high numbers of *B*s are present, as well as by some loss through mechanical inefficiency at meiosis and some infidelity in the non-disjunction process.

In *Locusta* and *Pantanga* species, as we have already indicated, there is frequently variation in the frequency of the *B*-chromosomes among and within the follicles of the testis. It arises from non-disjunction at mitosis. There is some evidence that, as a result of non-disjunction, cells without *B*-chromosomes contribute disproportionately fewer descendants to the germ line spermatocytes than cells which carry *B*s. There is, in other words, an element of somatic selection for *B*s (Kayano, 1971).

(c) *Structure and organization*

B-chromosomes are generally smaller than the *A*-chromosomes. Their behaviour at meiosis shows they are not homologous with *A*-chromosomes. This suggests a structural divergence from the *A*s which has been confirmed by comparisons of the chromomere patterns of *A*s and *B*s at pachytene of meiosis in rye and maize. *B*-chromosomes from the same population are generally similar in structure and their pairing behaviour at meiosis confirms their homology. Structural variants are by no means infrequent, however. An extreme case is found in *Aster ageratoides*. A standard form of *B* is accompanied by as many as twenty-four structural variants within a single population (Matsuda, 1970). The standard and morphological variants of the *B*-chromosomes in rye are shown in

Fig. 4.11. Polymorphism in respect of *B*s is common also in animal species. In *Melanoplus femur-rubrum*, for example, there is a large metacentric *B*, called B^{m} and two telocentrics, B^{1} and B^{t}, derived from it by breakage of the centromere.

The genetic organization of *B*-chromosomes is not clear. No Mendelian genes with major effects have been located on them other than those concerned with the control of non-disjunction, which we have described, and over the pairing of chromosomes in species hybrids (see p. 83). Neucleolus organizer regions are also conspicuously absent in *B*-chromosomes, although there is evidence of genes coding for ribosomal RNA in the *B*-chromosomes of rye (Flavell and Rimpau, 1975). The effects of *B*-chromosomes are much like those attributed to polygenic systems, producing a continuous rather than a discontinuous variation in the phenotype that is not readily distinguished from variation due to the environment. It was for this reason that they were initially considered to be inert, a view reinforced by the claim that *B*-chromosomes are frequently, but by no means always, heterochromatic (Jones, 1975).

The DNA composition of *B*-chromosomes, as revealed by biochemical analysis, does not differ in any significant way from that of the *A*-chromosomes in the three plant species so far investigated. In rye, maize and *Aegilops speltoides* the composition of the *B*-chromosome DNA is indistinguishable from that of the *A*-chromosomes, in respect of base ratio and the proportion of repetitive and non-repetitive sequences. The same applies to the grasshopper species *Myrmeleotettix maculatus* (Dover and Henderson, 1975). If the non-repetitive DNA in *B*s, like that of *A*s, embodies structural genes of major effect one must assume that the information they embody either is not transcribed or else that it merely duplicates to a large degree that contained elsewhere in the complement.

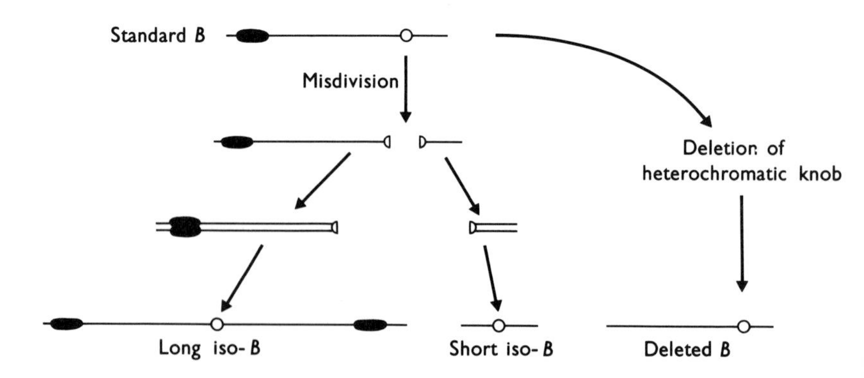

Fig. 4.11 *B*-chromosome polymorphism in rye, due to centromere misdivision and deletion of the standard fragment.

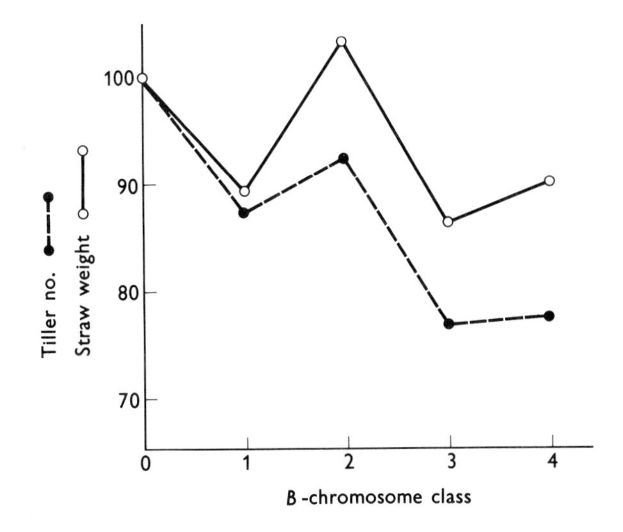

Fig. 4.12 The influence of *B*-chromosomes on vegetative characters in rye expressed as percentage of control (i.e. 0*B* plants). [Data from Müntzing, A. (1963) *Hereditas*, **49**, 371–426]

(d) *Effects*

Although most of the effects of *B*s upon the phenotype are of a quantitative rather than a qualitative nature, comparisons between large samples of individuals carrying different numbers of *B*s reveal that their effects are manifold and impinge on all stages of development. Fig. 4.12 and Table 4.9 show the effects of *B*s on morphological characters and fertility in rye plants. The general reduction in fertility and in vigour, as reflected by the morphological changes, with increase in *B* frequency is clear. We shall return later to consider the striking zig-zag pattern of variation in relation to odd and even numbers of *B*s in this species. The evidence from numberous other species conforms in demonstrating reduced growth and fertility associated with high numbers of *B*s, e.g. in the mealy bug, *Pseudococcus obscurus*. In passing it is worth noting that enhanced vigour is attributable to *B*s in low frequency in *Centaurea scabiosa* (Fröst, 1958). In

Table 4.9 Fertility, expressed as percentage seed set in rye plants with 0 to 8 *B*-chromosomes

B class	0	1	2	3	4	5	6	7	8
% seed set	49.5	31.4	34.2	21.5	5.1	7.1	1.7	0.1	0

view of the widespread distribution of *B*s among natural populations we might indeed expect that under certain circumstances they may enhance fitness.

Effects of *B*-chromosomes are equally clearly manifested at the cell level. The duration of the mitotic cycle in maize, *Lolium perenne* and rye is increased with the addition of *B*-chromosomes (see Fig. 4.23). Cell size in rye is also increased by the addition of *B*-chromosomes. In addition there is a substantial variation in the amount of RNA, and of protein due to *B*s (Fig. 4.13). It will be observed that total protein and RNA decrease overall in cells with high numbers of *B*s. The figure shows the same characteristic zig-zag relationship between phenotypic change and *B* frequency as applied to the morphological characters plotted in Fig. 4.12. The correspondence is not surprising because the morphological varia-

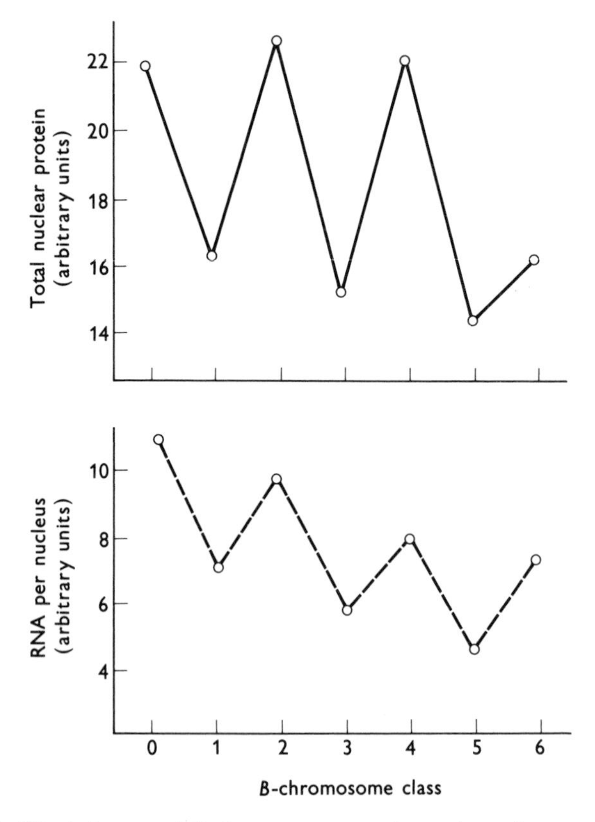

Fig. 4.13 The influence of *B*-chromosomes on the nuclear phenotype in rye. Quantitative assessments of total nuclear protein and RNA were made by microdensitometry after staining isolated root meristem nuclei with dinitrofluorobenzene (for total protein) and methyl green-pyronin *Y* (for RNA). [Data from Kirk, D. and Jones, R. N. (1970) *Chromosoma*, **31**, 241–54]

tion must, ultimately, be a reflection of variation in cell metabolism of the kind we have described.

(e) *Meiosis and recombination*

In 1956 Darlington suggested that *B*-chromosomes may serve to regulate the amount of variability among diploid populations. Support for this suggestion comes from work on rye by Moss (1966) who showed that the range of heritable variation for quantitative characters was greater among the progenies of plants with *B*s than those without *B*s. The increase in variability was over and above that due to the presence of *B*s among the progenies themselves. The control exerted by *B*s over the variability can be accounted for by their influence upon chiasma formation at meiosis.

In *Myrmeleotettix maculatus* the chiasma frequency of the *A*-chromosomes is increased in the presence of *B*-chromosomes. In spermatocytes without *B*s the chiasma frequency in one population averaged 14.13 per cell. With *B*s the frequency was 15.60 (John and Hewitt, 1965). It was shown also that the variation in chiasma frequency between cells and between bivalents within cells increased in the presence of *B*s, such that some cells and some bivalents had unduly high chiasma frequencies, others low. Both the effects on the mean and the variation in chiasma frequency could generate additional variability among gametes and progenies, the first by increasing the amount of recombination in general, the second by increasing the recombination in particular cells or particular chromosomes. It is the variation in chiasma frequency, both between and within pollen mother cells that is mainly affected by *B*-chromosomes in rye. They increase with increasing *B* frequency. It is worth emphasizing that in species like rye an increase in the chiasma frequency within a bivalent is accompanied also by a change in chiasma distribution. In rye bivalents with one or two chiasmata the chiasmata are restricted to the ends of the chromosomes. Additional chiasmata are formed interstitially. The effect therefore is not only to increase recombination generally but to generate recombination and novel recombinants from different regions of the chromosomes.

Figure 4.14 shows the effects of *B*-chromosomes on mean chiasma frequencies in maize, *Listera ovata* and in *Lolium perenne*. The *Lolium* results are of interest in showing that *B*s, while unquestionably controlling the chiasma frequency do so in this instance by depressing the number. This applies also in *Aegilops speltoides* (Zarchi *et al.*, 1972). While the effects of *B*s on chiasma formation and recombination vary somewhat from species to species there is no doubting their influence upon recombination and thereby upon the variability of gametes and of offspring.

(f) *Chromosome pairing in hybrids*

The chromosomes of *Lolium perenne*, a perennial outbreeding grass, are about 30% smaller and contain about 30% less DNA than the chromo-

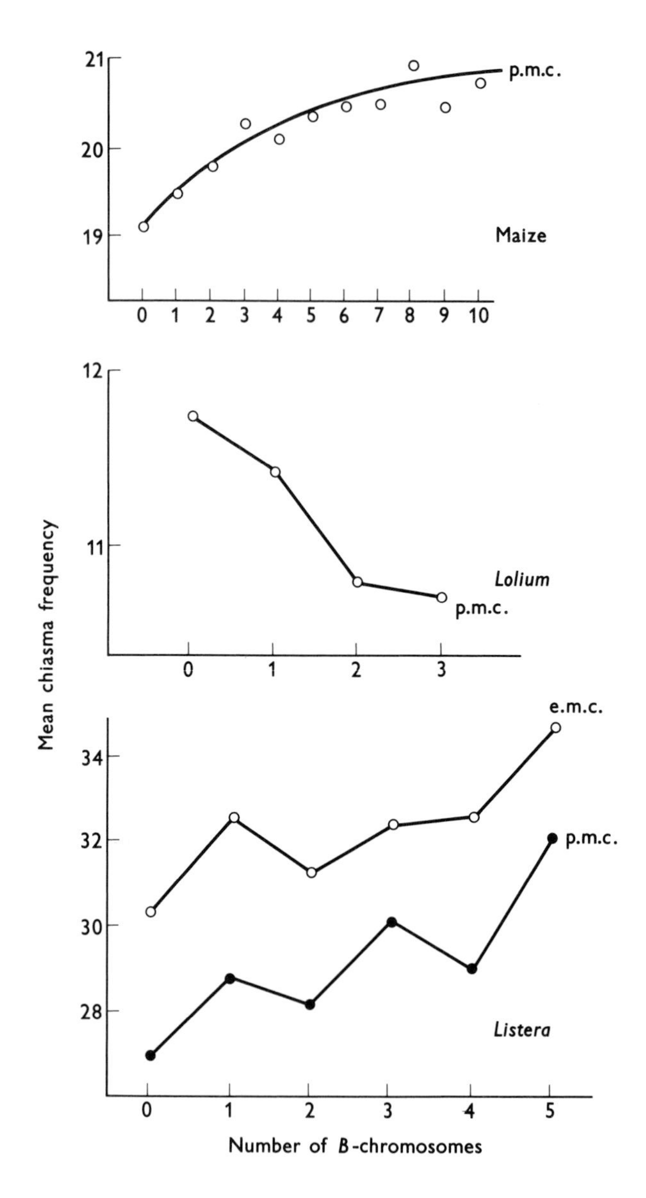

Fig. 4.14 The influence of *B*-chromosomes on the mean chiasma frequency of the *A*-chromosomes in pollen mother cells (p.m.c.) and in embryo sac mother cells (e.m.c.) in three species of flowering plants. [The maize data is taken from Ayonoadu, U. and Rees, H. (1968) *Genetica*, **39**, 75–81; the *Lolium* data from Cameron, F. M. and Rees H. (1967) *Heredity*, **22**, 446–450; and the *Listera* data from Vosa, C. G. and Barlow, P. W. (1972) *Caryologia*, **25**, 1–8]

somes of *L. temulentum*, an annual, inbreeding species of the same genus. Both are diploids with seven pairs of chromosomes. At meiosis in the diploid hybrid an average of 6, out of the possible 7, bivalents are formed in pollen mother cells. This is surprising because each bivalent comprises structurally different chromosomes, homoeologues rather than strict homologues. When the *B*-chromosomes of *L. perenne* are introduced into this F_1 diploid hybrid, however, there is a startling reduction in the bivalent frequency from six to four or fewer per cell. In tetraploids synthesized from the diploid hybrids the effect of the *B*s is even more dramatic (Fig. 4.15). Without *B*s there is pairing between both homoeologous and homologous chromosomes resulting in multivalent formation. The chromosome behaviour in fact is typical of that in an autotetraploid. When *B*s are present pairing is restricted to homologous pairs only, and all metaphase I configurations are bivalents as in an allotetraploid. That the bivalents result from the pairing of homologous chromosomes is established by their symmetry (Evans and Macefield, 1972). A similar 'diploidizing' effect of *B*-chromosomes has been reported in polyploid hybrids between the outbreeding diploids *Aegilops speltoides*, *Aegilops mutica* and the cultivated wheat *Triticum aestivum*. Hexaploid wheat normally behaves as an allopolyploid due to the presence of a gene on chromosome 5B which suppresses homoeologous pairing between chromosomes from the three constituent genomes (see p. 68). The *B*-chromosomes can substitute for the activity of this 5B gene in experimental situations where the 5B gene is absent (Dover and Riley, 1972).

(g) *Odds and evens*

One of the more puzzling aspects of the activity of *B*-chromosomes is the contrast between their effects upon the phenotype in odd and even numbers in certain plants. This is illustrated in respect of morphological and cytological characters in Figs. 4.12, 4.13, 4.14. The basis for such effects is not understood. A possibility put forward by Jones and Rees (1969) is that the *B*s may associate in pairs at interphase and that their activity in contiguous pairs differs from that as single chromosomes.

It will be recalled that the transmission of *B*s in many species is such as to generate an excess of gametes and progenies with even numbers. It is tempting to speculate therefore that the non-disjunction mechanism by which this is achieved is no mere fortuitous aberration but is an adaptive device ensuring a preponderance of individuals with even numbers of *B*s. The argument is strengthened by the fact that the effects of *B*s upon growth, in even numbers, are the least 'deleterious' (see Fig. 4.12).

(h) *Selection and adaptation*

Many of the phenotypic effects attributable to *B*-chromosomes are deleterious to growth and fertility. This is especially true for organisms carrying high numbers of *B*s. For this reason it has been argued that *B*s are nuclear 'parasites' and that their maintenance within populations is

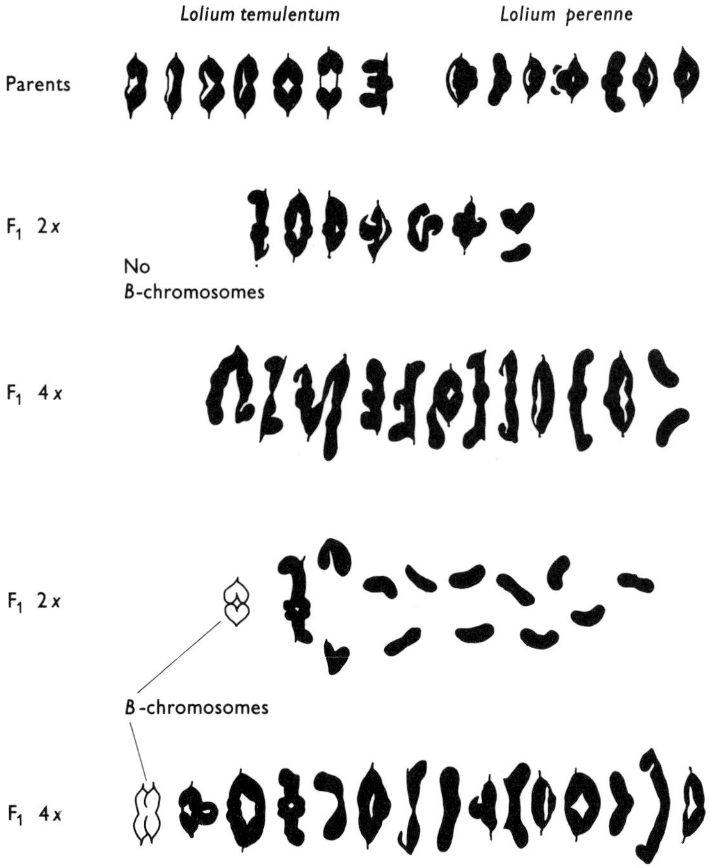

Fig. 4.15 *B*-chromosome control over chromosome pairing in *Lolium* species hybrids. In the absence of *B*s pairing takes place between *homoeologous* chromosomes in the F_1 diploid and derived tetraploid hybrids. In the presence of *B*s there is almost complete suppression of homoeologous pairing leading to bivalents only in the tetraploid. Pairing relationships in the hybrids are readily determined because of the size differences between the two parental complements.

solely dependent upon the mechanism of their transmission in heredity which tends to accumulate their numbers relative to the normal chromosomes of the complement. Yet *B*-chromosomes are widespread within many species and there are grounds for concluding that they confer superior fitness upon individuals and populations under certain experimental conditions.

(i) *Individuals*. *B* frequencies vary among populations within species. The variation, furthermore, is related to environment. Among populations of *Myrmeleotettix*, *Phleum* and *Centaurea* the *B* frequencies vary in relation to soil type or climate. In *Myrmeleotettix*, for example, populations in drier warmer areas have more *B*s than those in colder, wetter regions of Mid Wales (Hewitt, 1973). Such observations leave no doubt that the *B* frequency variation is dependent upon the environment. At the same time they do not establish that *B*-chromosomes enhance the fitness of individuals because one could argue that the frequency of *B*-chromosomes may increase within a population simply because the selection pressure is not sufficient to counter their accumulation during reproduction, and vice versa. There is, however, experimental evidence which shows that individuals carrying *B*-chromosomes display superior fitness in competition with individuals without *B*s under conditions of severe selection pressure.

(ii) *Selection in Lolium.* Seedlings from a population of *Lolium perenne* carrying *B*s were planted at different sites, one at sea level, the other at 900 feet (270 m); at different spacing, 6 inches (15 cm) and 2 feet (60 cm) apart at two seasons (spring and autumn). The mortality among seedlings varied, as would be expected, with the different 'treatments'. Of particular interest is the fact that the proportion of plants carrying *B*s among those surviving to maturity increased directly in relation to mortality. The conclusion is that under these conditions of increasing stress the fitness of individuals carrying *B*-chromosomes was superior to plants without *B*s (Rees and Hutchinson, 1973).

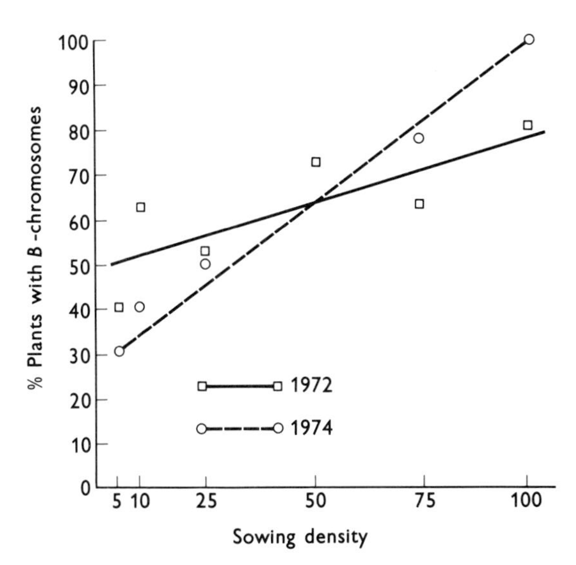

Fig. 4.16 The proportion of plants with *B*-chromosomes amongst survivors of *Lolium perenne* populations grown at different densities. The data from both years show the increase in the proportion of *B*s with increasing density. In 1974 only *B* plants survived in the high density populations in which the mortality is most severe. [Data from S. B. Teoh]

Further experiments in *Lolium* support this conclusion. Seeds from a population with *B*s were sown in small, 2 inch (5 cm) pots at densities varying from 5 to 100 per pot. Competition for nutrients etc. leads to increasing mortality with increasing sowing density. Fig. 4.16 shows the proportion of plants with *B*-chromosomes among survivors at the end of the first and third year. The proportion increases with increasing density of sowing. After three years as the graph shows, only plants with *B*-chromosomes survive the most severe selection pressures under conditions of high density. There is no doubt, therefore, that under certain conditions the *B*-chromosomes confer adaptive advantage. Fig. 4.16 is instructive, also, in showing that plants without *B*s in this experiment survive better than *B* plants in conditions of low sowing density. It will be observed that the proportion of *B* plants among survivors *drops* between 1972 and 1974 at a sowing density of five per pot. It is clearly important therefore to stress that the fitness in relation to *B*s is very much influenced by the particular environmental conditions which pertain.

(iii) *Populations*. *B*-chromosomes in most of the species investigated affect the frequency and distribution of chiasmata at meiosis. They must inevitably influence the amount and distribution of recombination and thereby the variability of offspring among populations. By these means they presumably influence the variability of the population as a whole and its capacity to adapt to a particular environment or to a changing environment. From the standpoint of selection and adaptation, therefore, the *B*-chromosome systems present a dual aspect. On the one hand they affect the fitness of the individual in terms of growth and development. On the other they affect the fitness of the population through their influence upon the variability made available for selection and adaptation.

(i) *Origin*

The way in which *B*-chromosomes originate is not known. In plants it is assumed they are derived from *A*-chromosome fragments which, subsequently, become structurally modified to a degree that conceals any residual homology with the chromosome ancestor. In many animal species the *B*-chromosomes resemble and show affinities with the *X*-chromosome (Hewitt and John, 1972). They often associate with the single *X* at prophase of meiosis although there is no case known of a chiasmate association. Even so the affinity between the *B* and the *X* is indicative of some degree of homology between the two with the implication that the *B* is a derivative of the *X*.

Amplification and deletion within chromosomes

The chromosomes of different species vary not only in number but also in size. Fig. 4.17 shows the complements of four diploid species of *Lathyrus*, each with 14 chromosomes. While the number is constant there is a 3-fold difference in total chromosome size between *L. hirsutus*, with

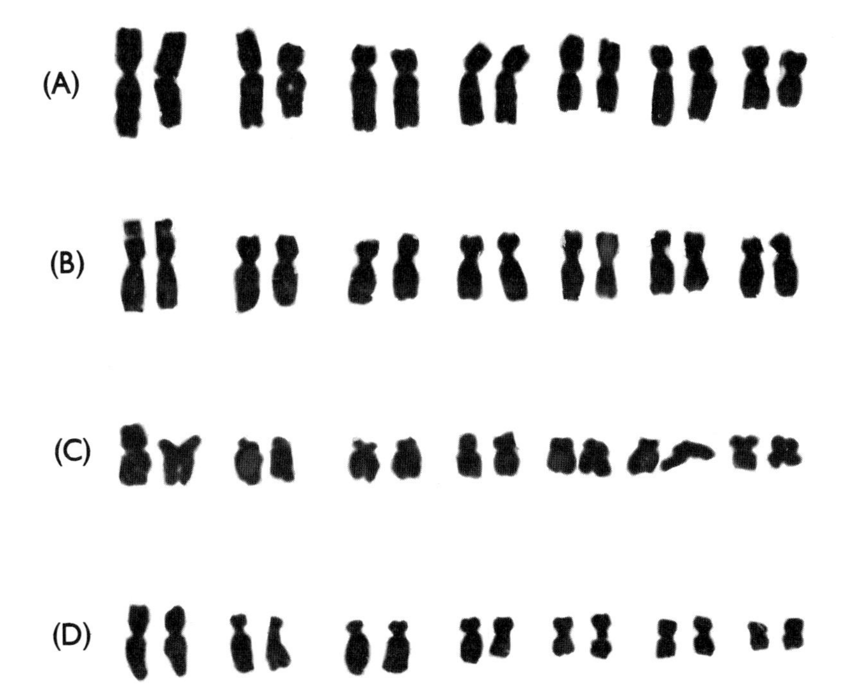

Fig. 4.17 The chromosome complements of four *Lathyrus* species. (A) *L. hirsutus*; A(B) *L. ringitanus*; (C) *L. articulatus*; and (D) *L. angulatus*.

the largest, and *L. angulatus* with the smallest chromosomes. The variation in size is very closely correlated with DNA content (Fig. 4.18). This survey in *Lathyrus* is typical of many which have established that substantial variation in nuclear DNA often accompanies the divergence and evolution of species, both in plant and animal genera. The magnitude of the DNA variation between species, aside from that due to polyploidy, is of a particularly high order in higher plants, in algae and in the amphibia (Fig. 4.2). In the higher plants for example, the diploid *Lilium longiflorum* has 100 times more DNA than *Linum usitatissimum* (see Rees, 1972). Even within the one family, the Ranunculaceae, there is an 80-fold variation. The variation among other groups such as the mammals and fishes is much less. Why this should be so is not known.

Two explanations have been put forward to explain the evolutionary changes in chromosomal DNA. The first is a *differential polynemy*, namely a change in the number of DNA strands within the chromosomes of different species (Fig. 4.19). It is now generally accepted, however, that the chromosomes, other than in specialized cells, are *mononemic* in all species and this proposal is therefore untenable. The second explanation is *lengthwise amplification and deletion* of segments to increase or decrease the

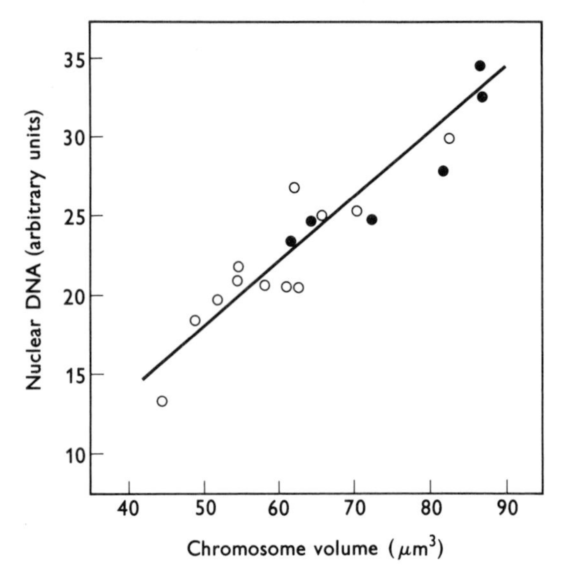

Fig. 4.18 The relationship between the mean nuclear DNA amount and total chromosome volume in 18 *Lathyrus* species. ●, perennial outbreeders; ○, annual inbreeders. [From Rees, H. and Hazarika, M. H. (1969) *Chromosomes Today*, **2**, 158–65]

amount of DNA within chromosomes (Fig. 4.19). The repetition of base sequences within the DNA of many species, to which we have referred earlier, is testimony enough to the capacity for amplification of chromosome segments. A substantial amplification and, for that matter, deletion of repeated sequences may be achieved during a short period of time. The 'bobbed' mutation in *Drosophila* is associated with loss of cistrons which code for ribosomal RNA in the *X* and *Y* chromosomes. The chromosome carrying the bobbed mutation may be restored to the wild-type condition by amplification of the rDNA cistrons to the number characteristic of the normal chromosome (Ritossa and Scala, 1969). In

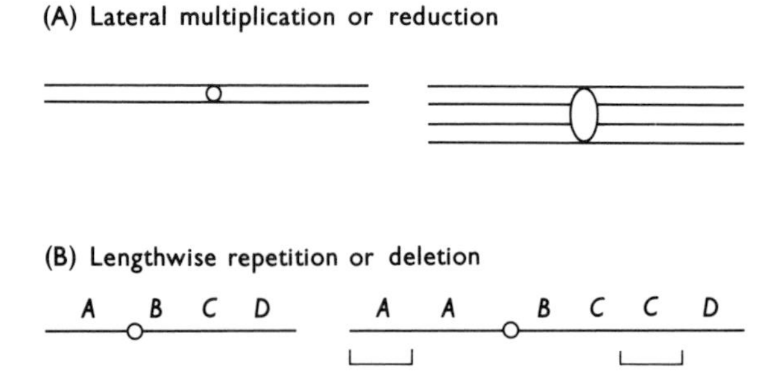

Fig. 4.19 Alternatives for quantitative change in chromosomal DNA.

Linum usitatissimum the number of rDNA cistrons and, indeed, of a large fraction of the DNA elsewhere in the chromosome complement is induced by specific environmental treatments (Timmis and Ingle, 1973; Evans, Durrant and Rees, 1966). Again, in a small proportion of cells in the hybrid *Nicotiana otophora* × *N. tabacum* a heterochromatic segment in one of the chromosomes is amplified to a prodigious extent to produce a chromosome 20 or 30 times the normal size and this during the span of one or a few interphases (Gerstel and Burns, 1966). Other evidence to confirm a lengthwise amplification as a cause for DNA increase within chromosomes comes from comparisons between related species with differing amounts of nuclear DNA.

(a) *Chironomus*

Chironomus thummi thummi has 27% more nuclear DNA than *Chironomus thummi piger*. The DNA content of certain of the bands in the polytene salivary gland chromosomes of *C. thummi thummi* is greater than in corresponding bands of *C. thummi piger*. The DNA difference between bands ranges from a factor of 2, to 4, 8, 12 and 16. The conclusion is that the higher DNA content of *C. thummi thummi* chromosomes results from localized lengthwise DNA repetition within bands. The series 2, 4, 8 etc. indicates a repeated duplication of the original bands or their copies (Keyl, 1965).

(b) *Allium*

The DNA content and the total chromosome volume in nuclei of *Allium cepa* are each 27% greater than in *A. fistulosum*. Despite the disparity in DNA amount the species hybridize readily. At meiosis in the F_1 hybrid all eight bivalents are asymmetrical, showing that the DNA changes associated with the divergence of the two species affected all chromosomes of the complement. Some bivalents, however, are more asymmetrical than others (Fig. 4.20). In some the difference in length between 'homologous' chromosomes at metaphase is no more than 10%. At the other extreme there is a 70% difference between the lengths of 'homologues'. At pachytene unpaired loops and overlaps range, as would be expected, with lengthwise repetition, from 10 to 70% of the length of the paired chromosome segments (Fig. 4.20 and Fig. 4.21). Similar loops and overlaps have been described in a *Lolium* species hybrid (Rees and Jones, 1967).

The number of loops or overlaps in each pachytene configuration is restricted to one or two. This suggests that amplification within chromosomes is highly localized. In the paired regions, furthermore, the pairing appears to be regular and complete which indicates that the chromosomes of *A. cepa* and *A. fistulosum* are structurally similar except for the localized regions of amplification in *A. cepa*. Restriction of amplification to certain regions of the chromosomes means either than such regions

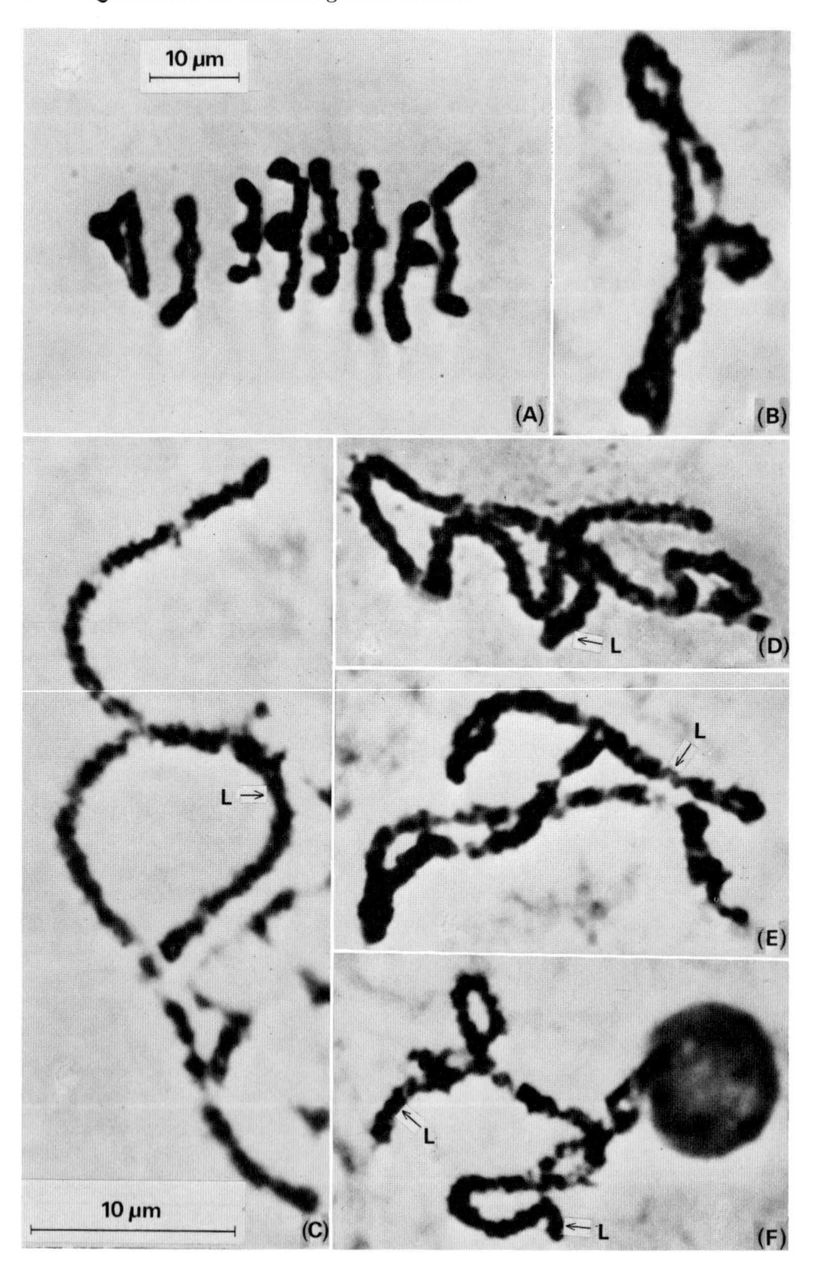

Fig. 4.20 Chromosome pairing in the F_1 hybrid between *Allium cepa* × *A. fistulosum*. (A) Metaphase 1 of meiosis showing 8 asymmetrical bivalents. (B) Diplotene with a pairing loop. (C)–(F) Pachytene bivalents with pairing loops (L). [From Jones, R. N. and Rees, H. (1968) *Heredity*, **23**, 591–605]

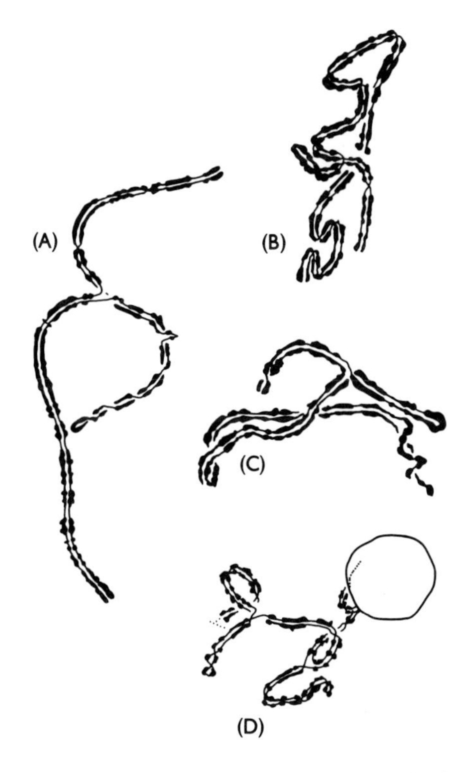

Fig. 4.21 The loops at pachytene in *Allium cepa* × *A. fistulosum*. (A), (B), (C) and (D) correspond with (C), (D), (E) and (F) in Fig. 4.20.

are prone to repetition or else that the consequences of repetition of these segments are tolerated whereas in other segments they are not. In either case the indications are that a special fraction of the chromosomal DNA is more likely to be amplified.

(c) *The location of quantitative DNA change*

Among *Lathyrus* species (Narayan and Rees, 1976) the amount of repetitive DNA is closely correlated with the total nuclear DNA amount (Fig. 4.22). The non-repetitive fraction does not vary significantly among species. We conclude that quantitative DNA variation during the evolution of these species is achieved through change in the amount of the repetitive component. It is generally assumed that the structural genes, i.e. those which are transcribed and translated into protein, are located mainly in the non-repetitive fraction. If this is so the variety of structural genes in *Lathyrus*, despite the massive alteration in DNA, remains relatively constant. The most obvious implication is that the change in total DNA concerns regulatory material but precisely what

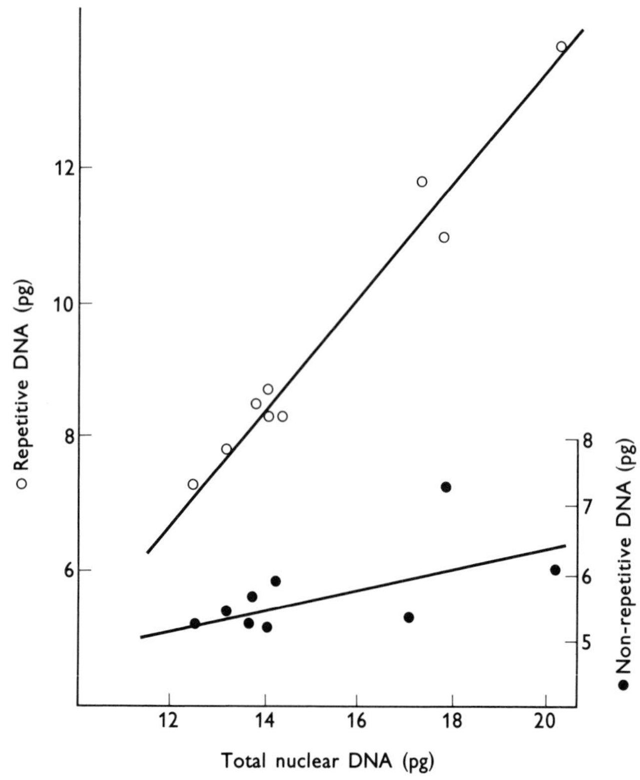

Fig. 4.22 Repetitive and non-repetitive DNA components in *Lathyrus* species plotted against the total nuclear DNA amount. [From Narayan, R. K. J. and Rees, H. (1976) *Chromosoma*, **54**, 141–54]

kind of regulation is involved is unknown. In *Lathyrus*, also, increase in DNA is associated with increase in the proportion of heterochromatin within the nucleus. We infer that the repetitive fraction is therefore located mainly in the heterochromatin. It will be recalled that prodigious DNA repetition is associated with centromeric heterochromatin in mammals and in *Drosophila* species. It will also be recalled that a heterochromatic segment in the *Nicotiana otophora* × *N. tabacum* hybrid is the site of rapid and extensive proliferation. This is not to say, of course, that DNA repetition is confined to heterochromatin. Indeed one of the earliest examples of repetition is that of the euchromatic *Bar* locus in *Drosophila* (Bridges, 1936). At the same time it is worth noting that a high degree of repetition is not necessarily reflected by a high heterochromatic content within nuclei. In wheat, for example, 60 to 80% of the DNA is repetitive, yet the amount of heterochromatin is very small indeed (Britten and Kohne, 1968; Smith and Flavell, 1975).

Fox (1971) finds in *Dermestes* beetles, as in *Lathyrus*, that DNA variation among species implicates mainly the repetitive fraction and, on a cytological basis, it is the heterochromatin which is increased disproportionately to euchromatin. Heterochromatin, it is frequently asserted, fulfils a regulatory role in gene activity but, as we have mentioned above, the mechanism of such regulation remains conjectural. One concludes, therefore, that the DNA variation among species appears, to a large extent, to be restricted to particular nuclear components. What this signifies, in physiological and in evolutionary terms, is by no means clear. It is worth, perhaps, suggesting that the answer may not be found in terms of one particular function but of many, embracing not only the activities of particular genes but the mechanics of chromosome organization and behaviour in general.

(d) *DNA loss*

Taking the broad view, the evolution of living organisms from microorganisms to higher plants and animals is accompanied by DNA increase. Yet within groups, both plant and animal, the evolution of (taxonomically) more advanced and specialized species is often accompanied by DNA loss. This is true for example of the bony fishes and of reptiles (Bachmann, Goin and Goin, 1972). Within plant groups also, the more advanced species often have less DNA than the more primitive, as in the genus *Crepis* and in *Lathyrus* (see Rees and Jones, 1972). In *Lathyrus* the loss is of the order of 60 % of the total nuclear DNA (Rees and Hazarika, 1969). In one sense this is surprising because the loss even of minute chromosome segments, induced by ionizing radiation for example, is more often than not lethal to the cell. An alternative explanation for the decreasing DNA amounts with specialization would be that DNA amounts had *increased* in older primitive forms while they remained more or less unchanged in the more advanced forms. Equally surprising are the spectacular increases in DNA associated with the evolution of groups such as the lungfishes, which, along with the urodele amphibia, have the highest DNA contents among eukaryotes (Morescalchi, 1973).

Ohno (1970) argues that sudden substantial increase in DNA marks the initiation of many evolutionary pathways; indeed that the duplication of genes, whether by polyploidy or repetition within chromosomes, is essential for the development of new, more complex forms. He maintains, for example, that DNA increase through polyploidy, among fish or amphibian ancestors was of crucial importance to the evolution of reptiles, birds and mammals. The basis of Ohno's compelling argument is that the evolution of a new gene from an existing one is achieved only by numerous alterations in the base composition at that locus. For this reason the function of one gene is lost before that of the new form is acquired. With duplication, one copy retains its functional efficiency while its copy accumulates those DNA base changes that ultimately specify the new product of what is now a new gene. This view, that major advances in evolution are dependent on a massive increase in DNA, is accompanied, paradoxically, by the argument that the 'redundant' component of the increased nuclear DNA is lost as species become more specialized. We have cited examples above of such apparent DNA loss in both plant and animal groups. Whatever the general pattern, however, the situation is by no means predictable or consistent. While in *Lathyrus* the more specialized, inbreeding species have more DNA than the more primitive outbreeders, in *Lolium* the more advanced inbreeders have more DNA than the outbreeding species (Rees and Jones, 1967). Inbreeding species in *Pinus* also have more DNA than outbreeders (Dhir and Miksche, 1974). In *Allium* the advanced groups of species with 14 and 18 chromosomes have more and less nuclear DNA respectively than the primitive group of *Allium* species with 16 chromosomes (Jones and Rees, 1968). Since we do not know precisely what kind of DNA is involved by these changes or how it functions there is no means by which we can evaluate the significance of DNA loss as opposed to gain during the divergence and evolution of new species.

4.2 DNA amount and phenotype

It will be recalled that polyploidy is frequently, although not invariably, identified with well-defined morphological and physiological modifications in the phenotype. In one sense this is surprising because the additional genetic material resulting from polyploidy varies in quality according to the species and, indeed, the genotype of the individual. The question arises, therefore, as to whether the addition of genetic material, of nuclear DNA *per se*, has certain inevitable effects upon the growth and development of the phenotype. If so we might well expect to find an effect on the duration of cell division on the naive assumption that the greater the nuclear DNA amount the longer the time required for the replication and division of the chromosomes.

DNA and duration of the mitotic cycle

In Fig. 4.23 the duration of the mitotic cycle in the root meristem of flowering plant species is plotted against the nuclear DNA amount. It is immediately apparent that the duration of the cycle increases with increasing DNA amount (by about 0.3 hours per picogram of DNA). This is despite the fact that the increase in DNA is achieved in a variety of ways, namely polyploidy and aneuploidy, the amplification of chromosome segments within diploid complements or the addition of *B*-chromosomes; in other words by the addition of DNA of widely differing quality and composition. This is not to say that the duration of the cycle is completely independent of DNA quality. Fig. 4.23 shows the mitotic cycles to be about 4 hours longer in dicotyledons than in monocotyledons with the same nuclear DNA amounts. Also, the rate of increase in the

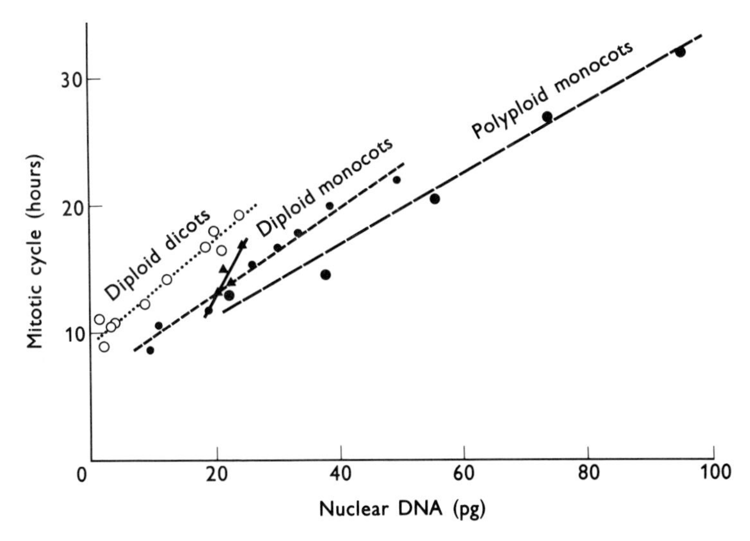

Fig. 4.23 Mitotic cycle times in root meristems at 20°C plotted against nuclear DNA amounts in flowering plants species. The black triangles are for rye plants with 1, 2, 3 and 4 *B*-chromosomes, in increasing order of cycle time. [From Rees, H. (1972) *Brookhaven Symp. in Biol.*, **23**, 394–418]

duration of the cycle is greater when due to the addition of *B*-chromosomes than of other chromosomes in the complement. These fluctuations, however, do not mask the overriding and highly predictable increase in cycle time with increasing DNA. As we have observed earlier there are equally well-defined correlations between the duration of meiosis and the nuclear DNA amount (Bennett, 1971).

DNA and cell size

Cell size in root meristems of flowering plants is also closely correlated with nuclear DNA amount. The increase in size follows DNA increase by both segmental amplification of chromosomes among diploids (Martin, 1966) and the addition of *B*-chromosomes.

Cell size and the duration of mitotic cycle together determine the rate of growth at the cell level. In tissues such as the root meristem both, as we have shown, are to a surprising extent dependent on the amount of DNA in the nucleus. There are indications that the relationship is of adaptive significance. Stebbins (1966) has pointed out that species of plants in temperate lands have larger chromosomes, and higher DNA content, than species in tropical regions. Bennett (1972) has demonstrated that short-lived species have a lower than average nuclear DNA content. The inference is that the restrictions upon such characters as the duration of mitotic cycles imposed by the nuclear DNA amount have a profound influence upon the distribution and evolution of species.

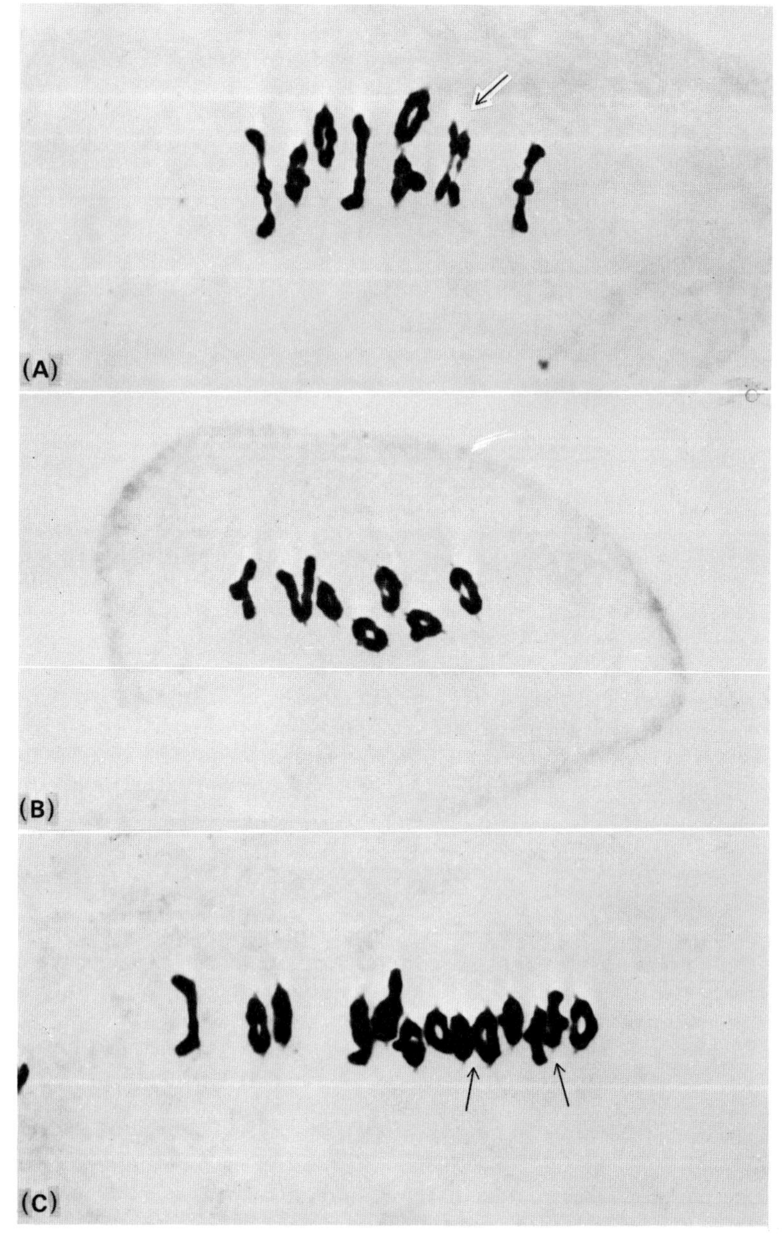

Fig. 4.24 Numerical chromosome change in rye. (A) First metaphase of meiosis with 4 *B*-chromosomes, as two bivalents (arrowed). (B) First metaphase in a trisomic, with trivalent. (C) First metaphase in an autotetraploid, 10 bivalents and 2 quadrivalents ($2n = 28$).

It is important in this context to bear in mind that the duration of mitotic cycles and the cell size differ very considerably as a normal consequence of growth and development between organs and tissues within the one individual. There is no doubt about the existence and operation of genetic mechanisms which control and regulate cell cycles and cell growth quite independently of nuclear DNA amount. In view of this it is perhaps even more surprising that the correlation between DNA amount, mitotic cycles and cell size among species is so consistent and predictable. It implies a crude and rigid constraint upon growth and development which appears to exert a profound influence upon the distribution and adaptation of species.

5

Qualitative change

Chromosome breakage, followed by the reunion of broken ends, is a normal and essential feature of crossing over at meiosis. Breakages of various kinds may also occur spontaneously during development. Their effects, like those of gene mutations, are unpredictable although usually lethal. They are lethal because most of the products of breakage are mechanically inefficient at mitosis. Acentric fragments, i.e. fragments lacking a centromere, are incapable of movement on to the spindle. Ring chromosomes and dicentrics resulting from breakage, with reunion of broken ends, are incapable of regular disjunction at anaphase. In each case the result is loss of genetic material. Mechanically efficient structural changes make up a small proportion of the products of spontaneous chromosome breakages. Even so they have important genetic consequences. The two most important are inversions and interchanges.

5.1 Inversions

Inversions are the result of two breaks within a chromosome followed by the reunion of broken ends in reverse order. When both breaks are in one arm the result is a *paracentric* inversion, a break in each of two arms gives a *pericentric* inversion. In either case a chromosome polymorphism may be generated with the inversion 'floating' within the population, comprised of inversion heterozygotes and inversion homozygotes alongside the structurally normal homozygotes. The frequency of each class will depend upon the relative fitness of the inversion in the structural heterozygotes and homozygotes relative to the normal genotype.

Because the two strands of the DNA helix are of different polarity the base sequences in the helix within the inverted segment are switched from one strand of the chromosome helix to the other. Such transfer does not in itself impair transcription. This was demonstrated, for example, within inverted chromosome segments in the *lac* operon region of *E. coli* (Hayes, 1968). It follows that transcription is selective in respect of the two strands within the DNA helix and also that sequences within the chromosome are transcribed from different directions. At the same time it must be supposed that where the inversion breaks occur within a unit of transcription the consequence would be to impair or prevent transcription. To some degree this may account for the lethal nature of many inversions in homozygous or hemizygous states.

Meiosis

At mitosis inversions do not impair the efficiency of transmission during cell division. Meiosis, also, is normal in the inversion homozygotes. In the heterozygotes, however, pairing between the normal and inversion chromosome is complicated and from the standpoint of fertility and recombination the consequences are profound.

(a) *Paracentric inversion heterozygotes*

At pachytene complete and effective pairing in the segment spanned by the inversion is achieved by the formation of a pairing loop (Fig. 5.1). The size of the loop, and hence the likelihood of chiasma formation within it, will depend upon the length of the inversion. If no chiasma forms within the loop, segregation is normal at both meiotic divisions. When a single chiasma forms within the loop the two chromatids involved form a dicentric bridge and an acentric fragment (Fig. 5.1 and Fig. 5.10). Loss of the fragment and random breakage of the bridge at first anaphase are the causes of deficiency in two of the gametes which in turn are inviable (in plants, generally) or incapable of producing viable offspring (in animals). The two chromatids not affected by chiasma formation in

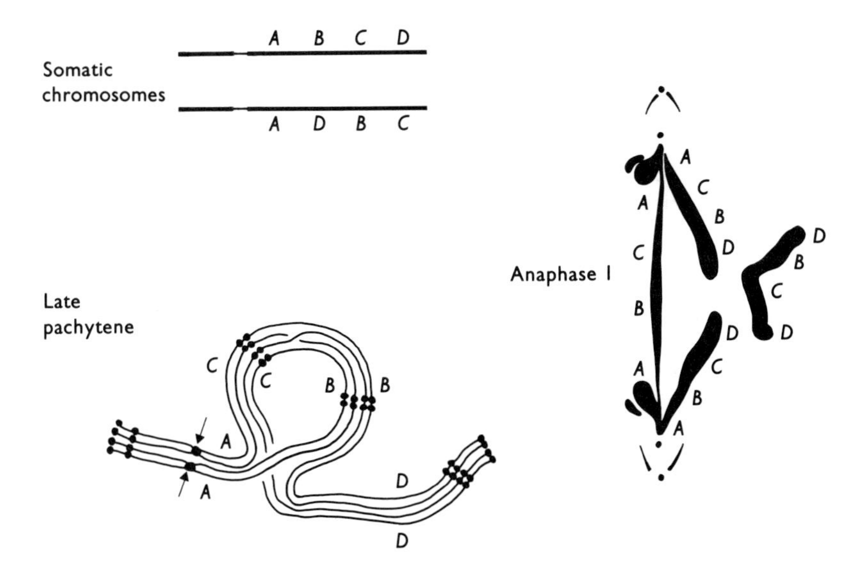

Fig. 5.1 Chiasma formation within the loop of a paracentric inversion heterozygote leading to bridge formation at anaphase, with acentric fragment. The chromatids not involved in chiasma formation are intact. Arrows indicate the centromeres.

the loop produce effective gametes. In each the gene sequence in the inversion segment will be of the non-recombinant, parental type. The result is that no recombinants for genes within the inverted segment appear among the offspring. The inversion thereby effectively suppresses

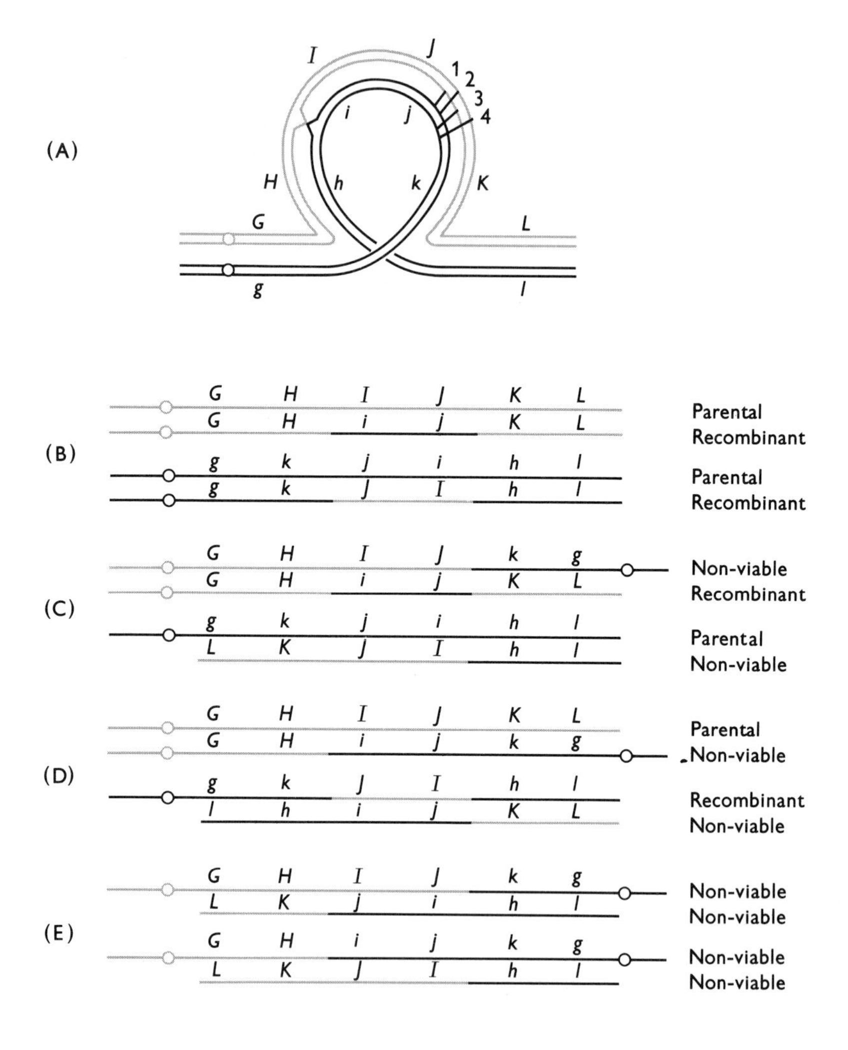

Fig. 5.2 The consequences of double cross-overs within the loop of a paracentric inversion (A). Assume the first cross-over to be constant, the second produces a 2-strand double cross-over (B), a 3-strand double (C) and (D) and a 4-strand double (E). Some viable recombinant gametes arise from the 2-strand and 3-strand doubles.

recombination throughout its length. The exception is where two-strand or three-strand double cross-overs take place within the loop. They are formed only when the inversion segment is a long one. Four-strand double cross-overs produce a double bridge and fragment and yield no effective gametes (Fig. 5.2). Leaving aside the other possible permutations of simultaneous chiasma formation within the loop and between the centromere and the loop it is sufficient to summarize the consequences of loop formation in the inversion heterozygote which are inviable or ineffective gametes and the suppression of recombination among genes in the inversion segment. Species of *Drosophila* and of related genera are exceptional in that inversion heterozygosity causes little or no reduction in fertility. There are two reasons for this: (1) there is no chiasma formation at meiosis in the male and (2) in the female the four nuclear products of meiosis are in linear order. The middle two nuclei are those involved in bridge formation at first anaphase and the functional egg nucleus is derived from one of the peripherals which carries no deficiency and is therefore viable and effective (Fig. 5.3). Thus there is effective suppression of recombination without inviability. This is one reason why paracentric inversions are particularly common in natural populations of *Drosophila*.

(b) *Pericentric inversions*

Crossing over within the loop of a pericentric inversion heterozygote does not result in bridge and fragment formation at anaphase I. Even so the gametes carrying recombinant chromatids are deficient for some

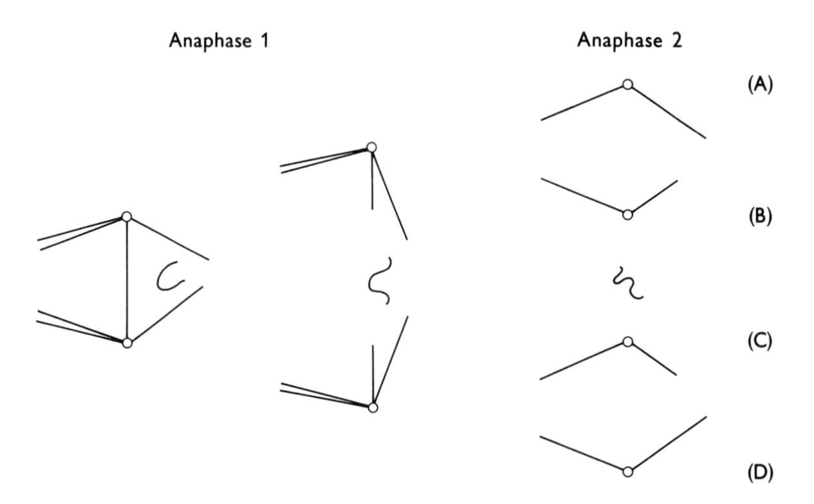

Fig. 5.3 Meiosis in the oocyte of *Drosophila* heterozygous for a paracentric inversion. The functional egg nucleus contains an intact, non-recombinant chromosome from either (A) or (D).

gene sequences while other sequences are duplicated (Fig. 5.4). As with paracentric inversions the consequences are inviable or ineffective gametes and the suppression of recombinants among the progenies for genes within the segment spanned by the inversion. However, pericentrics are particularly common among grasshoppers. The explanation is that no loops are formed in heterozygotes and, because pairing in the inverted segment is between non-homologous chromosome segments, no chiasmata are formed within it (White, 1973). It follows that recombination is suppressed but at no cost to fertility.

Distribution and selection

The effective suppression of recombination due to inversion implies that gene sequences within a segment spanned by inversion are maintained intact and segregate as one unit, a *supergene*, within a population. The frequency of an inversion within the population will therefore depend upon the relative fitness of individuals carrying these supergenes in homozygous or heterozygous combinations, bearing in mind that any advantage due to inversion must more than compensate for the infertility which, in most species, results from inversion heterozygosity. The widespread distribution of inversions in natural populations of both plants and animals testifies to their importance in adaptation and evolution. Among flowering plants inversions are reported in numerous species. In *Lolium perenne* 21 out of 50 populations investigated by Simonsen (1973) carried inversions. He also found (Simonsen, 1975) that 42 % of plants

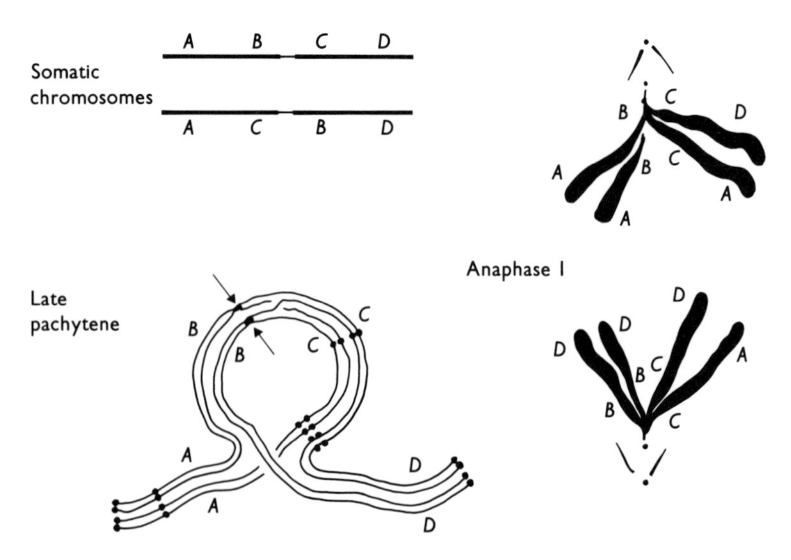

Fig. 5.4 Chiasma formation within the loop of a pericentric inversion heterozygote giving duplication and deficiency of segments in chromatids involved in chiasma formation. Arrows indicate centromeres.

within a population of a related grass, *Festuca pratensis*, were inversion heterozygotes. There is no information about the effects of selection upon the inversions in these or, indeed, in any plant species. In animals, in *Drosophila* and in grasshoppers particularly, there is a wealth of information.

The polytene nuclei of *Drosophila* larvae make the detection and identification of inversions relatively easy. Stone (1955) listed 592 different paracentric inversions in 42 species. He calculates that at least 6000 and possibly more than 30 000 inversions were associated with the evolution of the 650 species in this genus. Dobzhansky (1943) was the first to demonstrate the adaptive significance of inversions among natural populations. Chromosome III of *Drosophila pseudoobscura* in populations at Mount San Jacinto, California, is polymorphic for three inversion sequences, the Arrowhead (AR), the Chiracahua (CH) and the standard (ST) which is probably the 'wild type'. From 1939 to 1946 samples showed a consistent pattern of change within years. From March to June AR and CH increased at the expense of ST. From July to October ST increased while AR and CH declined in number. The conclusion was that selection favoured AR and CH gene sequences in the warm spring, ST in the hotter months of the year. The conclusion was confirmed by keeping flies in population cages. A sample from the wild population made up of ST/ST and CH/CH homozygotes and ST/CH heterozygotes were kept in cages at 25°C. The frequency of the ST chromosome, initially 11%, increased to 70%. At this point, with 70% ST and 30% AR chromosomes, the population attained equilibrium indicating that the CH chromosomes were maintained within the population due to the superiority of the heterozygote ST/CH over other genotypes (Dobzhansky, 1948). In cages kept at 16°C the relative frequencies of ST and CH remained unchanged to confirm that ST, as expected from the surveys of wild populations, is at an advantage only at high temperatures.

An inversion may of course be 'fixed' in a population in the homozygous form. In the *repleta* section of *Drosophila* 92 of 144 inversions in 46 species are fixed in this way (Wasserman, 1963). The remaining 52 'float' within populations which are polymorphic for inversion homozygotes and heterozygotes as in the *Drosophila pseudoobscura* stocks investigated by Dobzhansky.

In *Keyacris scurra* (formerly *Moraba scurra*), Australian populations are polymorphic for pericentric inversions located in two chromosomes (White, 1956; 1957a; 1957b; 1957c). The CD chromosome is found in one of four forms, *Standard*, *Blundell*, *Molonglo* and *Snowy*. The EF chromosome is found as *Standard* and *Tidbinbilla*. In both CD and EF forms the *Standard* chromosome is metacentric, the others are acrocentric or nearly so. Within populations, which produce one generation annually, the relative frequencies of the chromosome types are constant from year to year. Between populations, however, there is substantial variation indicating selection for particular sequences. In this species the effects of the inversion sequences upon the phenotype are directly recognizable. Adults homozygous for the *Standard* sequences are larger than *Blundell* and *Tidbinbilla* homozygotes. The gene sequences are additive in effect so that the heterozygotes are of intermediate size. In terms of viability, however, there is interaction. The inversion heterozygotes exhibit heterosis. Moreover, there is interaction between gene sequences on the CD and EF chromosomes. The least viable genotype is homozygous *Standard* for CD and homozygous *Tidbinbilla* for EF. Apart from demonstrating an adaptive polymorphism for pericentric inversions White's results are especially instructive in emphasizing that the effects of the gene sequences maintained by the inversions are manifold and complex. Changes in the frequencies of inversion sequences brought about by selection are therefore of a correspondingly complex nature both from the standpoint of gene action and interaction and the growth and development of the phenotype.

5.2 Interchanges

Interchange is the result of one break in each of two non-homologous chromosomes followed by re-union of the broken ends as in Fig. 5.5. The reciprocal exchange of segments involves no loss of genetic material, nor does it affect the efficiency of transmission of chromosomes at mitosis. In the interchange homozygotes meiosis is also normal, with regular bivalent formation and disjunction at anaphase of both divisions. There is consequently no cytological impediment to the production of viable gametes. In interchange heterozygotes, however, the exchange of segments between non-homologous chromosomes complicates the pairing at pachytene which, in turn, affects the subsequent progress of meiosis and of gamete formation.

Meiosis in interchange heterozygotes

Complete and effective pairing of all the homologous parts is possible only by association between the four chromosomes implicated in interchange, the two interchange chromosomes and the 'wild type' chromosomes from which they derive (Fig. 5.5). The fate of the interchange configuration and the nature of the gametes produced depend on the frequency and distribution of chiasma.

Consider the formation of four chiasmata, one at the distal end of each of the four arms of the association of four chromosomes, resulting in a ring. At first metaphase the ring may be orientated either as an 'open' or 'zig-zag' configuration. As an open ring the separation at first anaphase, although numerically and symmetrically disjunctional with two chromosomes moving to each pole, is non-disjunctional from the genic standpoint because the chromosomes at each pole are deficient for some chromosome segments and carry duplicates of others (Fig. 5.5 and Fig.

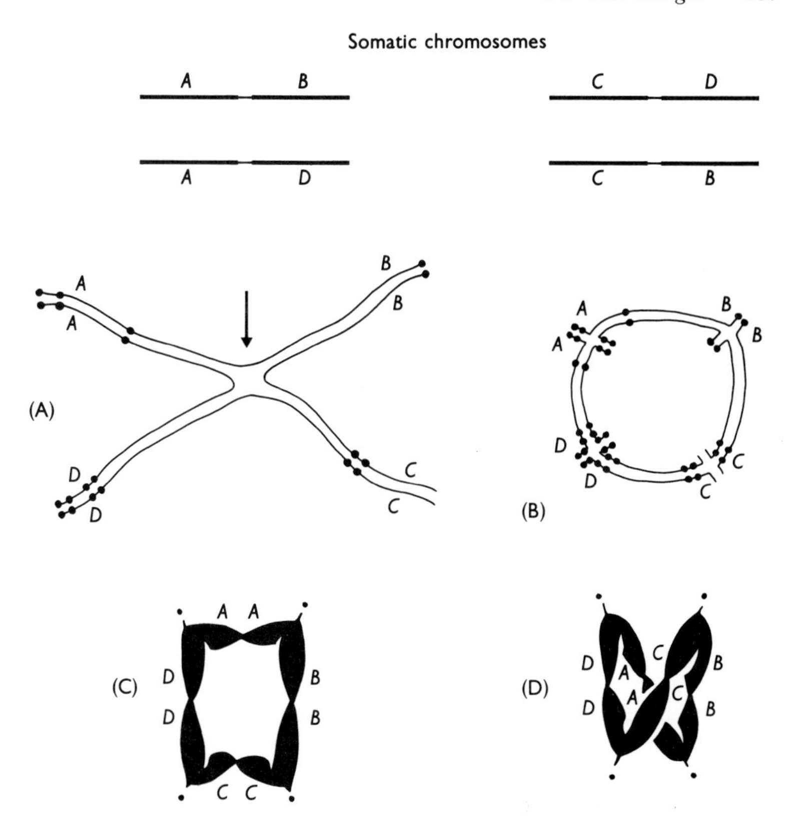

Fig. 5.5 Pachytene pairing (A) in an interchange heterozygote, followed by the formation of four terminal chiasmata (shown at diplotene, B). The open ring (C) at first metaphase leads to genetic non-disjunction, the zig-zag (D) to disjunctional separation. The arrow indicates the position of the centromere.

5.10). The gametes will be inviable or ineffective. If the ring orientation is zig-zag, disjunction is perfect and the gametes are viable and effective. The proportion of inviable gametes produced by the interchange heterozygote will depend on the proportion of cells in which the ring is orientated non-disjunctionally, as an 'open' configuration at first metaphase as opposed to a 'zig-zag', disjunctional orientation. While the theoretical expectation is 50% disjunctional orientation and 50% non-disjunctional, in many species, e.g. rye the orientation is disjunctional in about 65% of cells. It is worth noting that this proportion may be increased by selection with consequent improvement in the fertility (Thompson, 1956).

With three chiasmata in distal segments rather than four, a chain is produced. Again only a zig-zag arrangement gives disjunction and viable gametes. There is evidence to show that the proportion of chains which orientate disjunctionally is higher than of rings (Rees and Sun, 1965). It follows that a relatively low chiasma frequency improves fertility. If the distribution of chiasmata is such as to form two bivalents, there will be a 50% chance that their orientation is disjunctional.

If a chiasma forms between the centromere and the break point, i.e. in the *interstitial* segment and not distally as above in the *pairing* segment, half the gametes produced will be inviable even with a zig-zag orientation, i.e. with alternate centromeres moving to opposite poles. The additional infertility due to chiasmata in interstitial segments probably explains why interchanges are more common in populations of species in which chiasma formation is mainly distal in distribution.

Recombination

The chromosomes which segregate to opposite poles from zig-zag disjunctional arrangements at first metaphase are in two combinations, the original wild type pair *AB*, *CD* in Fig. 5.5 and the interchange pair *AD*, *CB*. Only gametes with these 'parental' combinations are effective. Any recombinants, *AB*/*AD*, *AB*/*CB*, *AD*/*CD* or *CD*/*CB* are inviable or else produce inviable zygotes. These are, of course, the products of non-disjunctional orientation. It follows that genes on the separate chromosomes, *AB* and *CD*, segregate together. The same is true for genes on *AD* and *CB*. Interchange therefore effectively prevents recombination between genes on different chromosomes by creating 'linkage' between them. The exceptions are genes near to the ends of the chromosomes where chiasmata may form. The suppression of recombination by interchange makes possible the maintenance of particular combinations of genes located on different chromosomes which, in theory, could be of adapative advantage in heterozygous or homozygous combinations.

Adaptation and selection

Paeonia californica is distributed along the California coastal strip from San Diego northwards to Monterey. Its centre of origin is at Ventura, about midway between San Diego and Monterey. In the Ventura region the populations comprise structurally normal diploids with 10 chromosomes which form 5 bivalents at meiosis. Towards both the southern and northern limits populations become increasingly heterozygous for interchanges (Walters, 1942). The distribution suggests an increasing adaptive advantage of interchange among the colonizing populations. Some contain individuals with four interchanges such that rings or chains are formed which involve all ten chromosomes at meiosis. A similar situation applies in *Oenothera*. The *hookeri* group in California form seven bivalents at meiosis. The *strigosa* and *biennis* groups, in inland eastern regions, are structurally heterozygous for interchanges forming rings or chains which, like some *Paeonia* populations, embrace all members of the complement. A common and important factor in both genera is that the interchanges are found in inbreeding, i.e. self-pollinating popula-

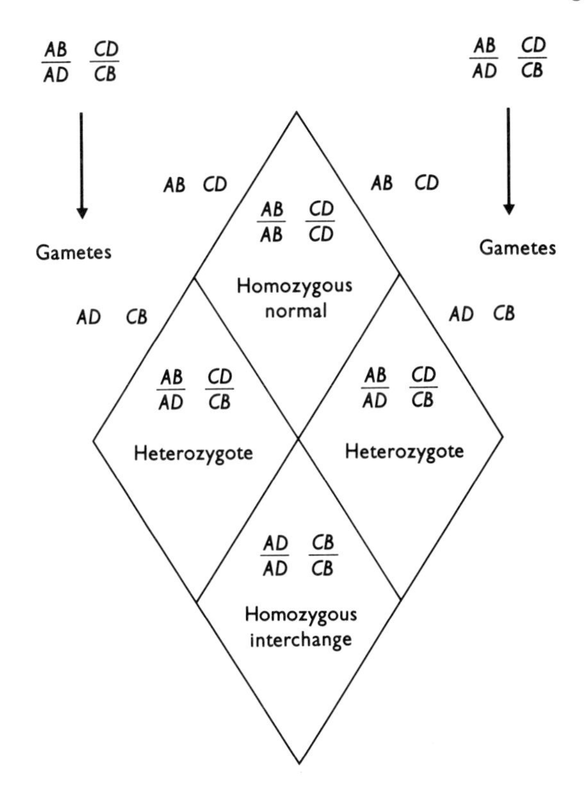

Fig. 5.6 The distribution of normal, interchange heterozygotes and homozygotes expected on selfing or crossing a single heterozygote.

tions as distinct from the bivalent forming ancestral populations which outbreed. The conclusion is that heterozygous gene sequences maintained by interchange are at an advantage under the conditions of enforced inbreeding due to isolation associated with migration and colonization. John and Lewis (1957) have described a comparable case in populations of the American Cockroach, *Periplaneta americana*. In coal mines in South Wales, where the isolation enforces inbreeding, interchange heterozygotes are common.

That the interchanges are favoured as conservers of particular heterozygous gene combinations with inbreeding is confirmed by experiment. The distribution of normal homozygotes, interchange heterozygotes and interchange homozygotes expected on selfing or intercrossing a single interchange heterozygote is in the ratio 1:2:1 (Fig. 5.6). On selfing there is an excess of interchange heterozygotes among the progenies in *Campanula persicifolia*, but not on intercrossing (Darlington and La Cour, 1950). In a rye stock with two identifiable rings of four due to two interchanges, *A* and *B*, the proportion of interchange heterozygotes among progenies increased with increasing generations of inbreeding. They outnumbered both homozygous classes by 2:1 (61 to 33). It is instructive that the *B* interchange heterozygotes showed no advantage over the homozygotes. Only some interchanges therefore—those which maintain particular gene sequences—are favoured. The extreme is found in the genus *Oenothera* where the interchanges are 'fixed' as heterozygotes and the homozygotes are lethal. Bearing in mind the infertility accompanying interchange heterozygosity the selective advantage of given gene sequences held together by interchange has to be substantial. We have indicated infertility may be reduced by heritable change in the frequency of disjunctional orientation of interchange configurations. In *Oenothera* species, for example, orientation of large rings or chains is disjunctional in 90% of cells.

Finally, although the experimental results show superiority only in respect of interchange heterozygotes, it may well be that interchange homozygotes may confer adaptive advantage over the normal wild type. In rye for example the different species are homozygous for different interchanges. *Secale montanum* and *S. cereale* differ by three interchanges. Each is homozygous for the interchange sequences. They are distinguishable by the ring and chain configurations at meiosis in the species hybrids.

Chromosome evolution and speciation

So far we have been discussing reciprocal interchanges between chromosomes which are relatively symmetrical and which lead to strictly qualitative changes in the organization of genetic material. Not all interchanges are of this kind. Some, particularly those involving acrocentric chromosomes which are common in the chromosome complements of animals, are extremely unequal and play a major role in chromosome and species evolution. They provide a mechanism for change in basic chromosome number and are the cause of quantitative as well as qualitative variation in genetic material.

One of the better known cases in plants is found in the genus *Crepis* (Tobgy, 1943). Tobgy hybridized *Crepis neglecta* ($2n=2x=8$) and *C. fuliginosa* ($2n=2x=6$) and found that at meiosis in the F_1 hybrid there was almost complete homology between the two parental sets of chromosomes. In many of the cells examined at first metaphase of meiosis he found two bivalents and an association of three chromosomes. Tobgy's interpretation of the chromosome changes associated with the evolution of *C. fuliginosa* from *C. neglecta* is presented in Fig. 5.7. As a result of an unequal interchange between the B and C chromosomes of *C. neglecta* one of the products, a small, 'inert' fragment is assumed to be dispensable. Its loss reduces the basic chromosome number from 4 to 3, as in *C. fuliginosa*. This is not to say that interchange is the only chromosome 'mutation' associated with the divergence of *C. fuliginosa* and *C. neglecta*. The chromosomes have diverged considerably in terms of size and DNA

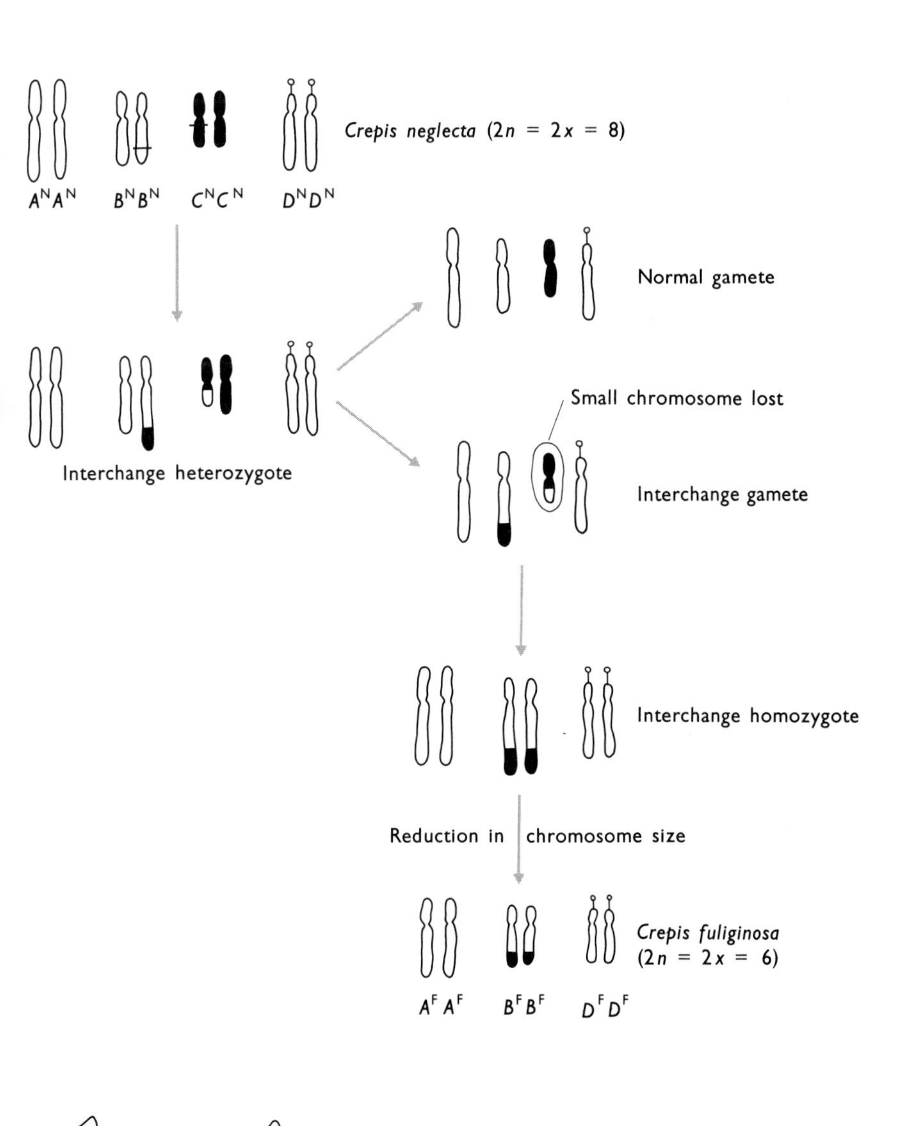

Fig. 5.7 The origin of *Crepis fuliginosa* from *Crepis neglecta* by unequal interchange and a reduction in the basic chromosome number. [After Tobgy, 1943]

content due to segmental amplification or deletion within the chromosomes (Jones and Brown, 1976).

In the *Bovoidea* the haploid number of chromosomes among the 50 or so species investigated ranges from $n=15$ to $n=30$. In the former the chromosomes are mainly metacentric, in the latter acrocentric, so that the number of chromosome *arms* is remarkably constant, from 29 to 30 in each haploid complement. For this reason it is postulated that the variation in chromosome number accompanying divergence and evolution of species within the group is, in the main, due to 'fusion' of the acrocentrics to produce metacentrics, with an accompanying reduction in the chromosome number. For example, in a species with $n=30$, made up of acrocentrics, reduction to $n=29$ is achieved by fusion of two acrocentrics to give one metacentric and 28 acrocentrics (Wurster and Benirschke, 1968). The phenomenon is called Robertsonian fusion (after Robertson, 1916). The mechanism could well be through unequal interchange of the kind described in *Crepis* above. On the other hand some workers claim that there is direct fusion of centromeres, in which case the chromosomes described as acrocentrics would in reality be telocentrics, i.e. with strictly terminal centromeres. Which of the two mechanisms applies is conjectural. White (1973) favours the view that reduction is by interchange and that true telocentrics are unstable and not, therefore, found as persistent members of chromosome complements. John and Hewitt (1968), on the other hand, claim that in some *Orthoptera* species at least there are stable telocentrics and instances of true centric fusion.

An instructive example of 'centric fusion' is described by Jones (1974) in the genus *Gibasis* (Commelinaceae). Two 'species' have, respectively,

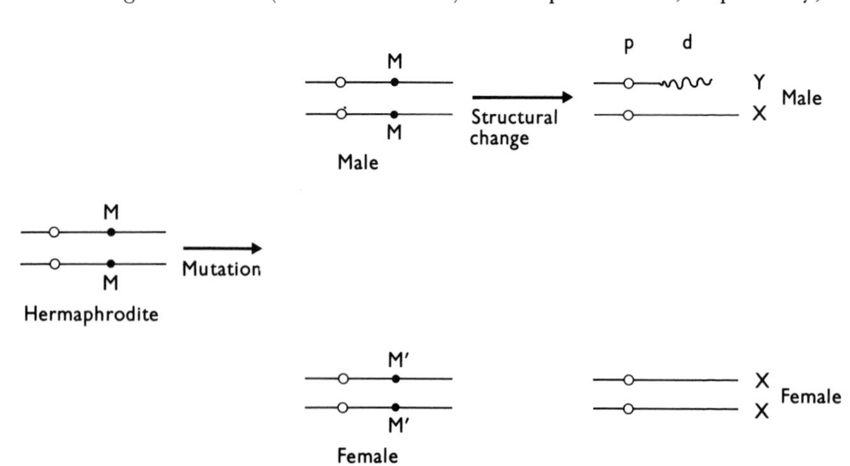

Fig. 5.8 Mutation, followed by structural divergence to account for sex chromosome polymorphism. The male is depicted as the heterogametic sex, i.e. producing gametes with different sex chromosomes. In many species it is the female that is heterogametic. p = pairing segment; d = differential segment.

$2n = 10$ and $2n = 16$ chromosomes, yet the latter is a tetraploid derivative of the former. The explanation is that the basic chromosome set of the diploid comprises 2 metacentrics and 3 acrocentrics. In the tetraploid, due to fusion, the basic set comprises 3 metacentrics and one acrocentric. Another example where polyploidy and centric fusion may, on the face of it, be confused, is in the *Salmonidae*. The chromosome numbers of species within this family correspond closely to the series $2n = 60$, 80, 100 and 120, strongly suggestive of polyploidy. Yet the number of chromosome arms in each species is relatively constant (c. 104) indicating Robertsonian fusion. The latter is confirmed by nuclear DNA measurements. The nuclear DNA amount in the sea trout (*Salmo trutta*, $2n = 80$) is similar to that of the Atlantic salmon (*Salmo salar*, $2n = 58$). Polyploidy is ruled out (Rees, 1964). Change in the basic chromosome number, as in the few of many examples reported, means change in the number of linkage groups. This in turn means change in the amount of recombination by independent segregation and the rate of release of genetic variability to gametes and progenies (see Chapter 6).

Sex chromosomes

The most widespread kind of chromosome polymorphism is that concerned with sex determination. Most animal species are *dioecious* (consist of unisexual individuals) and most species of higher plants are *monoecius* (hermaphrodite). In lower plants, such as mosses and liverworts sexual differentiation is more common (Mittwoch, 1967). It is widely accepted that mechanisms of sex determination evolved within monoecius species by gene mutation in such a way that the alleles segregate to specify different sexes (Fig. 5.8). Subsequent structural divergence, often involving diminution of the *differential* as distinct from the *pairing* segments of the chromosomes carrying the different alleles is the cause of the sex chromosome polymorphism characteristic of most dioecious species. The smaller of the sex chromosomes in mammals is the *Y*, in the red and white campions the *X*. In many species such as the grasshoppers the *Y* is absent.

Interchange between one of the sex chromosomes and an autosome often reinforces the chromosome divergence between the sexes. One such case is presented in Fig. 5.9. It accompanied the evolution of a number of species of both marsupial and placental mammals where males are invariably the heterogametic sex. Interchange between the *X* and an autosome is succeeded by loss of a small, presumably inert chromosome (Fig. 5.9). The final result is that the female is homozygous for a pair of interchange X/A chromosomes. The male is heterozygous for the interchange $X/A : A$ and also carries the *Y*. Consequently the male has one more chromosome than the female. In the marsupial *Potorus tridactylus*, for example, the male is $2n = 13$, the female $2n = 12$ (Sharman and Barber, 1952). Such a system has been described as XY_1Y_2/XX on the basis of trivalent formation in the male (Fig. 5.9), but as Ohno (1969) makes

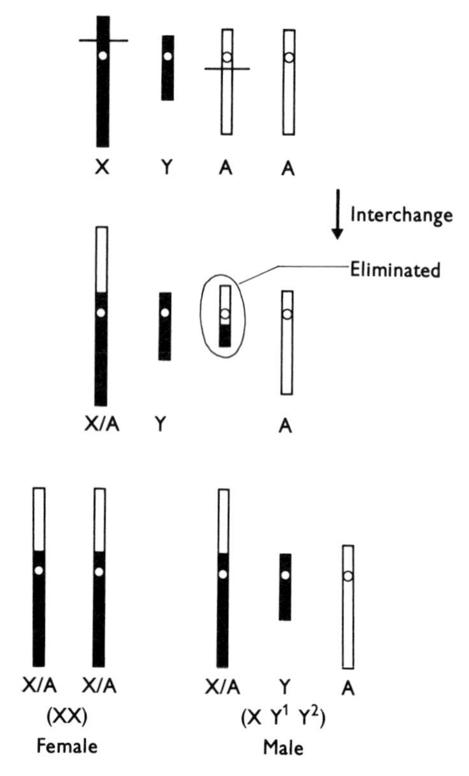

Fig. 5.9 Interchange between an X chromosome and an autosome (A), followed by loss of a small centric fragment. The consequence is that the chromosome number in the female is reduced by 1. Trivalent formation involving the X/A chromosome, the Y and the A in the male is the reason why the system has been described as an XX/XY^1Y^2 mechanism. [After Ohno, 1969]

clear such a classification is unnecessarily confusing in that it implies the development of a new sex determination mechanism whereas the mechanism determining sex is unchanged. It is simply that one of the sex chromosomes happens to be implicated in interchange. Comparable but often more elaborate interchanges involving sex chromosomes are by no means uncommon (Mittwoch, 1967; Ohno, 1969). The significance of such changes is readily explained. We must suppose that the complex of genes embraced by the sex chromosome and a particular member of an autosomal pair confers superior fitness and that the complex is maintained intact by interchange. Evidence to support this view comes from Barlow and Wiens (1975). In an East African mistletoe (*Viscum hildebrandtii*) all the males are heterozygous for one or more interchange. The number of interchanges associated with the sex chromosome, moreover, increases among populations from south to north, precisely as would be

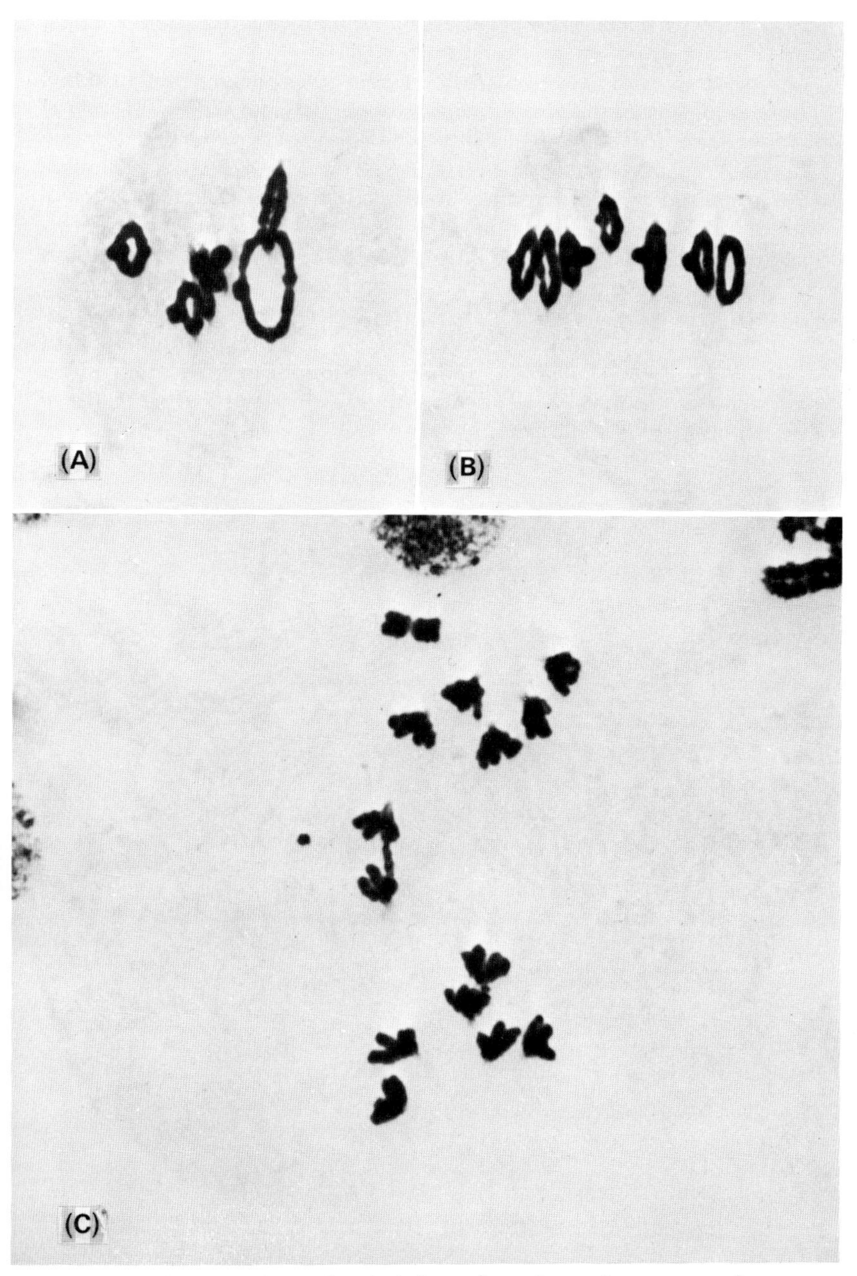

Fig. 5.10 First metaphase of meiosis in an interchange heterozygote in rye; (A) with an 'open' ring, (B) a zig-zag ring of 4. (C) First anaphase in a rye plant heterozygous for a paracentric inversion.

expected with selection for particular interchange combinations adapted to different environmental conditions.

Finally, we wish to draw attention to one other phenomenon in many, but not all species showing sex chromosome polymorphism. It is that of 'compensation'. In the mammals the male has one active *X* and a largely heterochromatic, relatively inert *Y*. As we describe elsewhere one of the *X*-chromosomes in the female is 'switched off' such that transcription of sex chromosome genes is similar in the two sexes. In *Drosophila* 'compensation' is not achieved by switching off one of the two *X*s in the female but is imposed, in the main at least, by 'compensation' genes (regulators), located within the *X* itself, which depress the activity of other *X*-chromosome genes in the double *X* females or in any individual with more than one *X* (Muller and Kaplan, 1966). Polymorphism in relation to sex chromosomes is often elaborate not only in terms of structure but also of metabolism.

6

The regulation of recombination

6.1 Genetic systems and the states of variability

Heritable variation derives in the first instance from mutation of the genetic material. Mutation, however, is but one of many factors contributing to the heritable variation within a population. Consider, for example, two genotypes *AAbb* and *aaBB*. Let us assume that the alleles *A* and *B* have similar effects upon the phenotype and, likewise, that the effects of *a* and *b* are similar to one another. The situation is typical of genes which make up a polygenic complex. The result would be that *AAbb* and *aaBB* would be of similar phenotype. By intercrossing we get F_1 heterozygotes *AaBb* and by further intercrossing or selfing a range of genotypes and phenotypes in the F_2. In the parental forms *AAbb* and *aaBB* the variability is of a potential kind (a *homozygotic potential* state) (Fig. 6.1). It is converted to the *heterozygotic potential* by intercrossing and, subsequently, to *free variability*, as in *AABB* and *aabb* genotypes, by recombination at meiosis in the *AaBb* heterozygotes (see Mather, 1973). The example suffices to illustrate that three ingredients: (1) the initial mutations at *A* and *B*; (2) the breeding system in relation to intercrossing; and (3) recombination at meiosis, contribute to the release of heritable variation among the F_2 phenotypes. The three together constitute what Darlington (1958) has called the *genetic systems* of populations. It was Darlington who perceived that adjustments in any one of these components serve to regulate the amount and kind of heritable variation within populations and, moreover, that the adjustments in themselves could be adaptive in controlling the variability made available for selection.

6.2 The flow of variability

In organisms which reproduce exclusively by asexual means the only source of variability is mutation. The same is true for sexual inbreeders. Because of their homozygosity, chiasma formation in these organisms yields no recombinants. Among the homozygotes, as in *AAbb* and *aaBB* above potential variability persists in the absence of hybridization. In an outbreeding population, in contrast, there is a continual flow of variability due to intercrossing and recombination, from the free to the

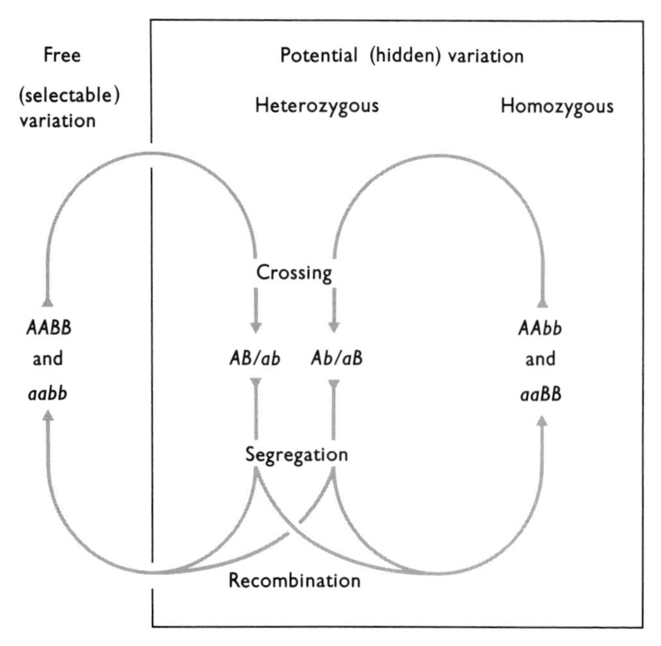

Fig. 6.1 The states of variability. The model assumes the effects of *A* and *B* upon the phenotype to be similar. *a* and *b* are also assumed to have the same effect on the phenotype. Free variability is expressed as phenotypic differences among genotypes (e.g. *AABB* and *aabb*) and may be fixed by selection. The homozygous potential variability is concealed (c.f. *AAbb* and *aaBB* which have similar phenotypes) and is therefore not discriminated by selection. The homozygous potential is converted to the heterozygous potential by crossing and, subsequently, to free variability following recombination. The amount of recombination determines the rate of flow from the heterozygous to the free state. The arrows show the direction and causes governing the flow of variability from the one state to the other. [From Mather, K. (1973) *Genetical Structure of Populations*. Chapman & Hall, London]

potential state, from the potential to the free; from the homozygotic to the heterozygotic potential and vice versa (Fig. 6.1). The *rate* of flow of variability even within the outbreeding populations is, however, subject to regulation, imposed by the amount and distribution of recombination at meiosis. The aim of this chapter is to indicate briefly how such regulation is achieved and to illustrate some of the consequences.

6.3 The regulation of recombination by structural and numerical changes

Structural changes

For genes located on different chromosomes segregation is independent such that gametes from the heterozygote *AaBb* are *AB*, *Ab*, *aB*, *ab* in equal numbers with the recombinants making up half the total. *Interchange* is one means whereby the recombination may be effectively suppressed. Likewise, *inversion* suppresses recombination between genes on the same chromosome (Chapter 5). A consequence of both kinds of structural change is to conserve particular gene combinations which may confer superior fitness in particular environments.

Numerical changes

Numerical changes of two kinds serve, like inversion and interchange, to restrict the flow of variability by reducing recombination. The one is allopolyploidy which, on account of a restriction of pairing to homologous and genetically identical chromosomes at meiosis, produces gametes similar to the two parental genotypes (Chapter 4). The second is a reduction in the basic chromosome number. The consequence is a smaller number of genes which segregate independently. The extreme is found in *Ascaris megalocephala* where all the genes are linked to one another on the single chromosome in the complement. Grant (1959) reports that short-lived species of higher plants have, on average, lower basic chromosome numbers than the longer-lived species. He argues that the difference is adaptive, on the following grounds. For short-lived species the demand is for uniformity so that the offspring, which have to re-establish the population in each generation, are similar to the parental forms adapted to that environment. Reducing the basic number and thereby the number of genes which segregate freely and independently is one way of promoting uniformity. Conversely, an increase in the basic number fosters a freer flow of variability through greater recombination.

Autopolyploidy, simply by increasing the number of genes and of chromosomes, increases the number of possible gene combinations and thereby the range of variation. There is one other kind of numerical change which affects recombination and the variability of offspring and populations, namely *B*-chromosomes (Chapter 4). Almost invariably they increase the amount of recombination among the chromosome complement as a whole or in particular chromosomes or chromosome segments. Either way *B*-chromosomes promote a rapid rate of flow of variability.

It will be observed that the regulation of recombination by structural and numerical chromosome changes is, for the most part, an 'all or nothing' situation. With interchange or inversion the recombination is completely suppressed among the chromosomes or chromosome segments affected. There is little or no element of fine adjustment. The regulation imposed by the action of genes upon recombination, on the other hand, allows for such adjustment.

6.4 Genotypic control

We have emphasized earlier that the chromosomes have a dual aspect, that of genotype and that of phenotype and that the latter is influenced not only by changes in cell environment imposed by differentiation and by the external environment, but also by the action of genes which are borne by the chromosomes themselves. The control imposed by genes over the form and behaviour of the chromosomes is aptly named *genotypic control*. We shall consider here those aspects of genotypic control which impinge upon recombination.

In *Neurospora crassa* a number of *recombination genes* regulate the frequency of recombination within other genes or between genes within short segments of the chromosomes (Catcheside, 1977). One of these, a recessive gene *rec-1*, controls the frequency of recombination at the *his-1* locus which is located on the same chromosome but at some distance from *rec-1*. In the homozygote *rec-1*, *rec-1* recombination within the *his-1* locus is increased ten-fold above normal. It is characteristic of these recombination genes in *Neurospora* that they influence recombination in specific chromosome segments at some distance away.

A single recessive gene, *as*, on chromosome 1 of maize causes almost complete failure of chiasma formation throughout the whole chromosome complement (Beadle, 1930). A gene in *Hypocheris radicata* causes failure of chiasma formation in one particular chromosome, chromosome IV (Parker, 1975). Heterochromatic segments which segregate as Mendelian genes increase chiasma frequencies in a number of grasshopper species (John, 1973). In others they reduce the frequency (D. I. Southern, 1970). These are but a few of numerous examples of genes at a single locus causing variation in chiasma frequency and in chiasma distribution. There are instances where the control is exerted by many genes, i.e. by polygenic systems, as shown for example by investigations of inbred lines.

Chiasma frequency in inbred rye

Secale cereale ($n = 14$) is an outbreeder. Inbred lines may be produced by enforced self-fertilization. The lines are homozygous, but as a result of segregation during inbreeding, homozygous for different gene combinations. Table 6.1 shows the mean chiasma frequencies in four such lines. The differences between means are significant and heritable showing that the segregation during inbreeding involved genes determining the frequency of chiasmata in the pollen mother cells.

The chiasma frequencies in all lines are substantially lower than in normal, heterozygous rye plants. Homozygosity evidently is a cause of inbreeding depression in respect of chiasma formation as for other more familiar aspects of the phenotype such as plant height and grain yield. There is a parallel also in the *heterosis* displayed in F_1 heterozygotes derived from crosses between the inbred lines. The F_1 chiasma frequencies are substantially higher (Table 6.1).

That the control over the chiasma frequency is polygenic is evident from the continuous nature of the variation among inbred lines and F_1s. The consequences upon the regulation of recombination are, therefore,

Table 6.1 The mean chiasma frequency per pollen mother cell in 4 inbred lines of rye and their F_1 hybrids. [From Rees, H. and Thompson, J. B. (1956) *Heredity*, **10**, 409–24]

Lines and hybrids	Chiasma frequency
P3	12.95
P6	11.97
P12	11.90
P13	12.11
P3 × P6	14.21
P3 × P12	14.49
P3 × P13	14.63
P6 × P12	14.31
P6 × P13	14.25
P12 × P13	14.48

very different from those imposed by structural change which allow for no fine adjustments such as may be achieved through genotypic control.

Chiasma distribution in inbred rye

In bivalents with one or two chiasmata, the chiasmata are located distally, near the chromosome ends. In bivalents with three or more chiasmata the extra ones are formed interstitially (Fig. 6.2). An increase in the bivalent chiasma frequency in rye, as in many other species, is therefore accompanied by a redistribution of chiasmata within chromosomes. In terms of recombination it means that in bivalents with two or fewer chiasmata the gene sequences at all but the ends of the chromosomes are unaffected. With three or more chiasmata gene sequences in the middle regions of the chromosomes are vulnerable to disruption by crossing over. Bivalents with three or more chiasmata are of course more numerous in plants with high chiasma frequencies. In these not only is there more recombination but recombination in different parts of the chromosome complement.

A redistribution of chiasmata both between bivalents and within bivalents may also be achieved without change in frequency. Jones (1967) has described a rye line with a relatively low chiasma frequency, about 13 per cell, but which has an unusually high frequency of bivalents with interstitial chiasmata (Fig. 6.3).

The examples suffice to establish the capacity for heritable adjustments in the frequency and distribution of chiasmata and of recombination. That control over recombination affects the variability of progenies and of populations is also established in both experimental and natural populations.

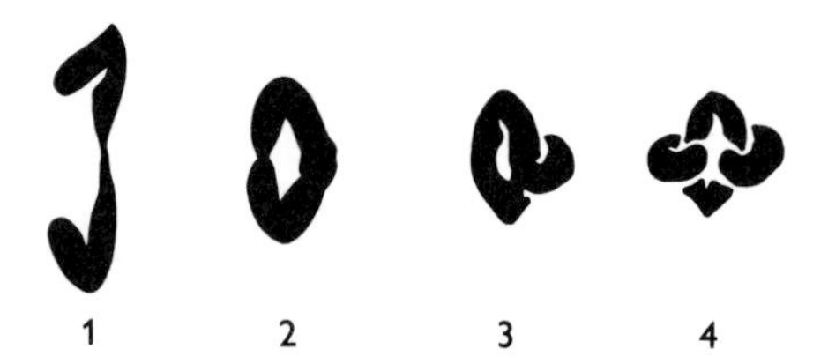

Fig. 6.2 Rye bivalents at first metaphase of meiosis with 1, 2, 3 and 4 chiasmata. Note the change in chiasma distribution with changing frequency. All chiasmata are distal in bivalents with a frequency of 1 and 2. Interstitial chismata are found in bivalents with 3 or more.

6.5 Recombination and variability

McPhee and Robertson (1970) crossed two strains of *Drosophila melanogaster* differing in the average number of sternopleural bristles, a character under polygenic control. The F_1 hybrids were of two kinds, one being heterozygous for inversions in chromosomes II and III, the other 'normal', *i.e.* without inversions. The range of variation in bristle number among the F_2 progenies of F_1 inversion heterozygotes was very much less than among F_2s from hybrids without inversions. This is precisely as expected. In the former there is effective suppression of recombinants and of the release of variability among progenies. The consequences of suppressing recombination were further manifested by the results of selection for high and low bristle numbers among F_2s and their derivatives. Measured as the difference in bristle number between

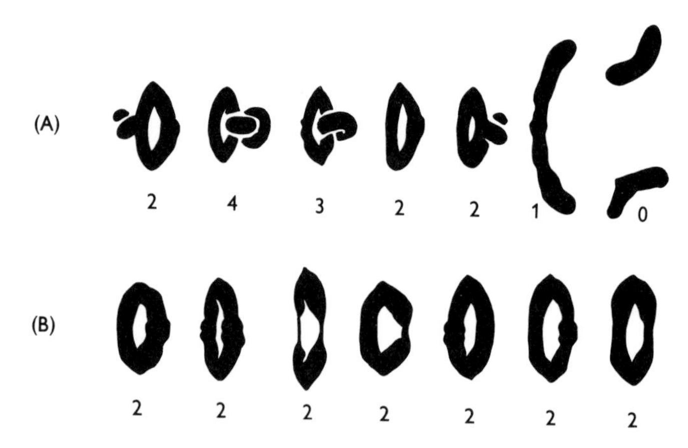

Fig. 6.3 First metaphase of meiosis in pollen mother cells of two rye genotypes with a similar chiasma frequency (14) but differing in chiasma distribution. In (A) there is wide variation in the number of chiasmata among bivalents within the cell (from 0 to 4). (B) is more characteristic of the species with a uniform distribution of chiasmata amongst the bivalents. [After Jones, G. H. (1967) *Chromosoma*, **22**, 69–90]

high and low lines the selection was only 25% as effective among derivatives of inversion heterozygotes as compared with lines derived from F_1 hybrids without inversions.

Moss (1966) has shown how the redistribution of chiasmata, brought about by *B*-chromosome action, influences recombination and variability in experimental populations of rye. In this species, as already mentioned, the *B*-chromosomes increase the frequency of chiasmata in the interstitial segments of the chromosomes. The localized increase in recombination is reflected by an increased variability among the progenies of parents with *B*s as compared with those without *B*s (Chapter 4).

6.6 Chiasmata and variability in natural populations

That the regulation of recombination, exercised by genotypic control over chiasma frequency and distribution, influences the variability within natural populations in an adaptive way is established from surveys among populations of *Lolium* and *Festuca* grasses (Rees and Dale, 1974).

Variation

Within *Lolium perenne*, *L. multiflorum* and *Festuca pratensis*, all diploids with 14 chromosomes, there is a significant variation in the mean chiasma frequencies among natural populations. The variation is heritable. As in other species of Gramineae the chiasmata are localized in terminal regions of the chromosomes when the chiasma frequency is low, spreading to interstitial regions as the frequency increases.

Variability and chiasma frequency

The variability between plants within populations for a number of characters such as flowering time is lower in populations with high chiasma frequencies and vice versa (Fig. 6.5). This relationship, which is consistent throughout all three species, is at first glance at variance with the experimental results of McPhee and Robertson and of Moss to which we refer above. The contradiction is only apparent. As shown in Fig. 6.4 the consequence of high chiasma frequencies is, initially, to release potential variability and thereby to expose the free variability to selection. The effect of selection in turn, is the fixation of some genotypes at the expense of the dissipation of the remainder and, as a result, an overall reduction in the variability, both free and potential within the population. By contrast, in populations with low chiasma frequencies much of the variability is preserved, protected from disruption and release by the absence or rarity of interstitial chiasmata. Gene sequences within each chromosome segregate virtually as single units in heredity, as supergenes. Since each chromosome type within the complement may be allelic for a number of such supergenes it follows, also, that the population is genically highly polymorphic and, in comparison with populations with high chiasma frequencies, highly variable.

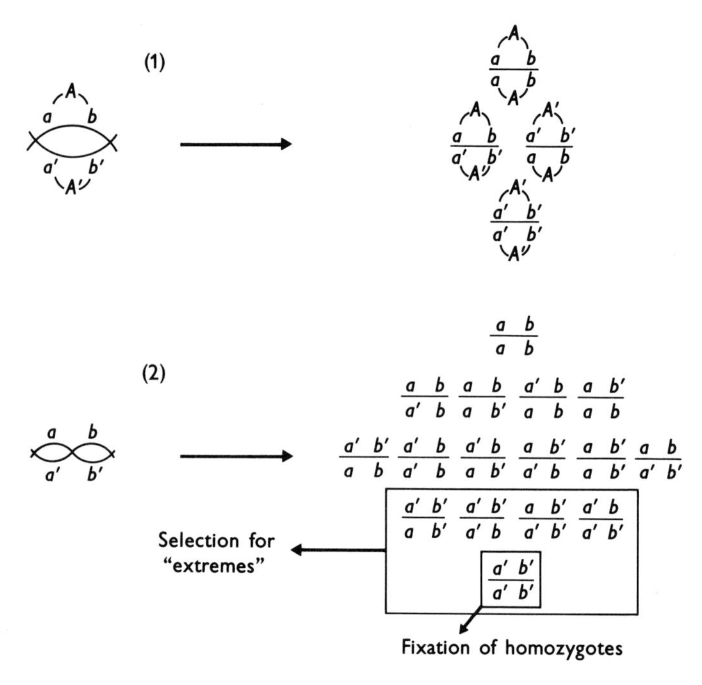

Fig. 6.4 Model showing how chiasma frequencies and distal localization (1) hold together supergene sequences, *A* and *A'*. Components of the supergene are *ab* and *a'b'*. In (2) interstitial chiasmata and high chiasma frequencies have fragmented the supergenes and released the variability. Part of the variability is fixed by selection, the remainder is dissipated. [From Rees, H. and Dale, P. J. (1974) *Chromosoma*, **47**, 335–51]

Chiasmata and selection

Selection for early and later flowering among eight populations of *Festuca pratensis* showed selection to be more effective among populations with low chiasma frequencies. This we expect on the grounds that, in the short term, the response to selection would be rapid through reassortment of highly diverse chromosomes; diverse because they segregate as supergenes protected from disruption in the absence or rarity of chiasmata and, in the long term, by the release of potential variation following the occasional formation of chiasmata in interstitial segments. In populations with high chiasma frequencies the variability is already largely dissipated with the result that the response to selection is restricted. In *Lolium perenne*, as in *F. pratensis*, selection is more effective in populations with low as compared with high chiasma frequencies.

Chiasmata and specialization

Short-lived populations of *Lolium perenne* and *L. multiflorum* have higher chiasma frequencies than more perennial populations from which they

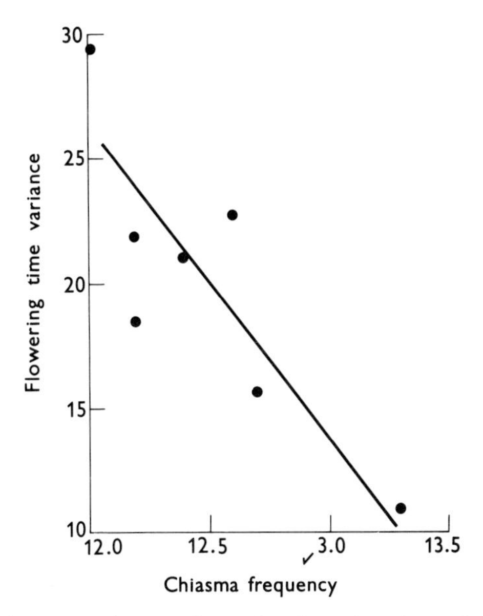

Fig. 6.5 The between-plant variance for flowering time plotted against the mean chiasma frequency in populations of *Lolium multiflorum* (From Rees, H. and Dale, P. J. (1974) Chromosoma, **47**, 335–51).

have evolved (Fig. 6.6). The same applies in rye (Sun and Rees, 1964). Intensively bred strains of *Lolium perenne* have higher chiasma frequencies than populations in 'natural pastures'. The model in Fig. 6.4 accounts for the high chiasma frequencies in these specialized forms. It is through recombination, particularly in interstitial segments, that new variability may be generated. In these populations the high chiasma frequency is a relic of the recombination essential for their evolution. To confirm this interpretation it would be expected that intensive selection for specialized forms would be accompanied by, albeit unconscious, selection for high chiasma frequencies. This precisely was the finding of Harinarayana and Murty (1971). Selection for late and early flowering within populations of *Brassica campestris* was accompanied by an increase in chiasma frequency.

Chiasmata, fitness and flexibility

The situation in the *Loliums* and in *Festuca* illustrates how increasing the chiasma frequency, especially within interstitial regions of the chromosomes where chiasmata are normally infrequent, generates new forms by recombination that may be fixed by selection imposed by nature or by the breeder. In the short term this facilitates rapid adaptation and high fitness. The very means which provides for this rapid adjustment,

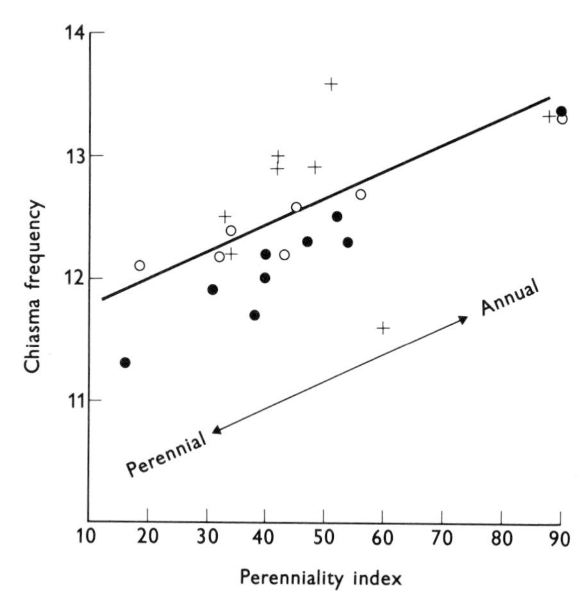

Fig. 6.6 Longevity (Perenniality index) plotted against the mean pollen mother cell chiasma frequency in populations of *Lolium perenne* (●), *L. multiflorum* (○), and *Festuca pratensis* (+). The perenniality index is the percentage of plants flowering during the first season after sowing. [From Rees, H. and Dale P. J. (1974) *Chromosoma*, **47**, 335–51]

the increase in recombination, is, however, instrumental in dissipating the variability such that the population has a lower capacity to adapt to further demands imposed by selection in a changing environment. Fitness is therefore at the expense of flexibility. As Darlington and Mather (1949) have emphasized the requirements for fitness in the short term and flexibility in the long term are often in conflict. In terms of chiasmata the conflict may be resolved by compromise. In most natural populations of the *Lolium* and *Festuca* species, for example, the chiasmata are mainly restricted to distal segments such that a variety of gene sequences, relatively well adapted to the environment, are maintained intact within the population. At the same time, occasional chiasmata in interstitial segments release a limited amount of variability such that there is the capacity for some flexibility in the face of changing selection pressures.

We have restricted our comments to the evidence for the regulation of variability imposed by genotypic control over recombination. In a wider context we should take account of the distribution and the mode of action and interaction of the genes which determine the variability. Moreover, we have focused attention on populations within a few species of grasses. The reason is that little information is available from other sources. Even though it be limited, the information from *Lolium* and *Festuca* nevertheless provides strong evidence of the adaptive consequences of adjustments in chiasmata upon the variability of populations and their capacity to respond to selection.

7

Recombination at mitosis

Recombination is a characteristic of meiosis. The extent of recombination between particular genes allows us to 'map' the chromosomes, i.e. to locate the genes within the chromosome complement. The methods are well established and familiar. In diploid organisms the mapping is achieved by scoring the proportion of phenotypes arising by recombination among F_2 or backcross progenies which, in turn, reflects the distribution of recombination among chromosomes at meiosis in the F_1 hybrids. In haploids, such as the fungi, the genotype of the gametes can be established directly and yields the same information. That recombination may also take place during mitosis was first established by Stern (1936) in *Drosophila*. The phenomenon has since been investigated in detail in *Aspergillus* by Pontecorvo (1953) and Pontecorvo and Kafer (1958), and is fully discussed in a companion volume of this series (Catcheside, 1976). More recently it has been used to provide information about the location and distribution of genes in the human chromosome complement.

7.1 Drosophila

Females of *D. melanogaster* of the genotype $\frac{+y}{sn+}$, heterozygous for the two recessive genes *y* (yellow body) and *sn* (singed bristles) on the *X*-chromosome are normally wild type in appearance. Stern observed in some cases 'twin spots' of yellow and singed on the body and also spots of yellow or singed alone. He ruled out mutation on the grounds that twin spots (which would require two simultaneous mutations) were more common than single spots. His explanation was that crossing over took place during interphase of mitosis, following chromosome duplication, to give twin or single spots as shown in Fig. 7.1. The size of any one spot would depend on the amount of cell multiplication following the crossing over. That twin spots were more numerous than yellow and singed is precisely as expected from the distribution of *y* and *sn* on the *X*-chromosome. Twin spots are due to a cross-over between the centromere and *sn*, yellows to a cross-over between *sn* and *y* and singed spots to a

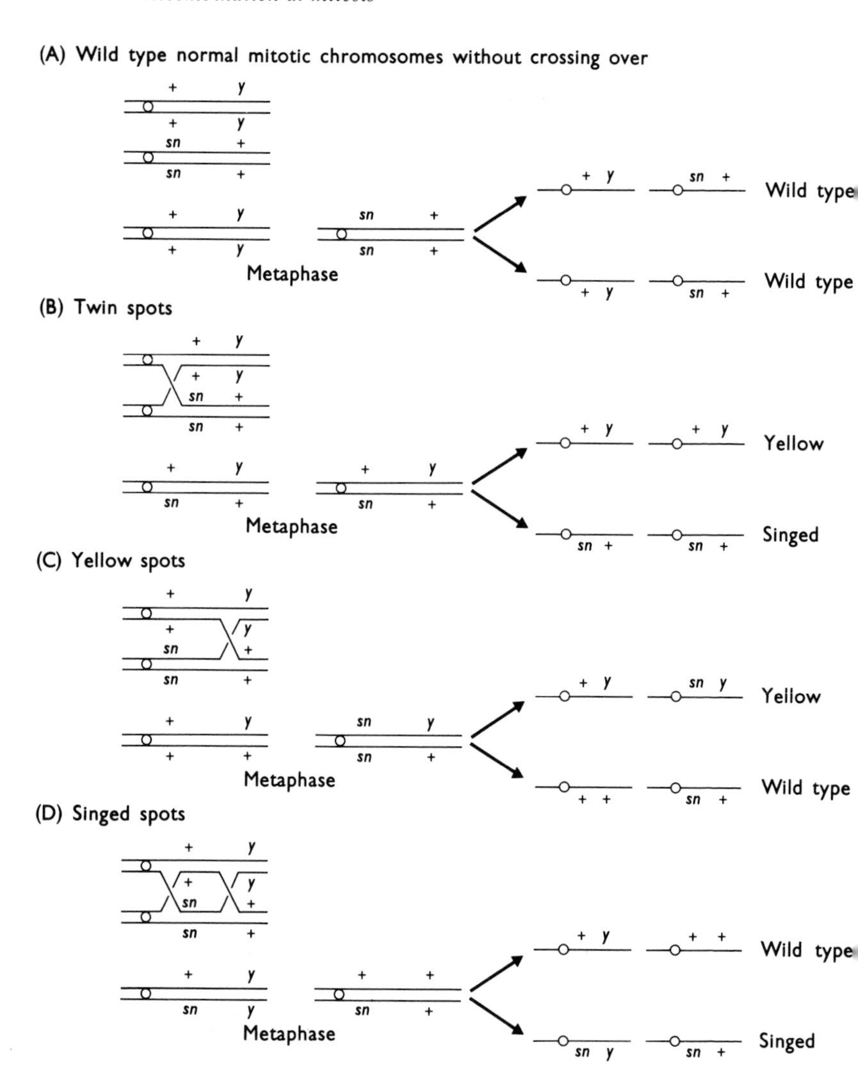

Fig. 7.1 Somatic crossing over in *Drosophila*. (A) a normal mitotic division; (B) crossing over between the centromere and the *sn* locus can give rise, assuming the metaphase orientation shown, to adjacent daughter cells homozygous for *y* and *sn*. Subsequent cell multiplication gives the twin spots of *yellow* and *singed* tissue. (C) Crossing over between *sn* and *y* gives spots of tissue with the yellow phenotype and in (D) a double cross-over gives spots with the singed phenotype.

double cross-over. There was evidence that crossing over at mitosis was genetically controlled. It was enhanced, for example, in the presence of the gene *minute*. In 1970 Stern demonstrated crossing over at mitosis between mutants *w* and w^{65}, within the white eye locus, also located in the *X*-chromosome of *Drosophila*. In 4 out of 6137 female heterozygotes, $\frac{w+}{+w^{65}}$, the intragenic crossing over was detectable as red, wild type, patches within the white eyes. Mutation was ruled out because no trace of pigmentation was found among 27 557 white eyed homozygotes and hemizygotes (males) used as controls. While Stern's work in *Drosophila* serves to demonstrate the occurrence of mitotic crossing over, the detailed analysis of the process comes from the investigations of Pontecorvo and his colleagues in the ascomycete *Aspergillus nidulans*.

7.2 Parasexuality in aspergillus

Wild type *Aspergillus* grows on minimal agar and produces dark green asexual spores, conidia. Nutritional (auxotrophic) mutants, which are unable to grow on minimal agar, are available to mark all eight chromosomes of the haploid complement. Strains which carry a spore colour mutation as well as a nutritional mutation are particularly useful in analysis, as is explained below.

Heterokaryon formation

When yellow and green conidia from two auxotrophic strains are mixed together, some of the hyphae within the mycelium will contain nuclei from both strains, as a result of fusion. These are heterokaryons as distinct from the homokaryons which contain one kind of nucleus only. The heterokaryon is readily isolated by growing the mycelium on minimal agar. In the heterokaryon one nucleus complements for the nutritional deficiency in the other and vice versa. The result is that the heterokaryon grows successfully, the homokaryons do not and are thereby screened out. During asexual reproduction the heterokaryon produces chains of conidia. All conidia in each chain contain one or other of the two nuclei. For this reason the homokaryons are readily recovered.

Aspergillus nidulans also reproduces sexually. In this case there is nuclear fusion as well as hyphal fusion prior to meiosis and ascus formation. Pontecorvo showed, however, that nuclear fusion in heterokaryons may occur independently of this sexual process. Such fusions are rare (1 in 10^6). The diploid products may be recovered as hyphae from sectors with exclusively green conidia within the mycelium of the heterokaryon; green because green is dominant to yellow in the diploid nucleus. Alternatively the diploids may be obtained by germinating conidia on minimal agar. Only the diploid conidia will thrive by virtue of complementation (Fig. 7.2).

Segregation and recombination

Diploid colonies of *Aspergillus* are unstable. When grown on a complete

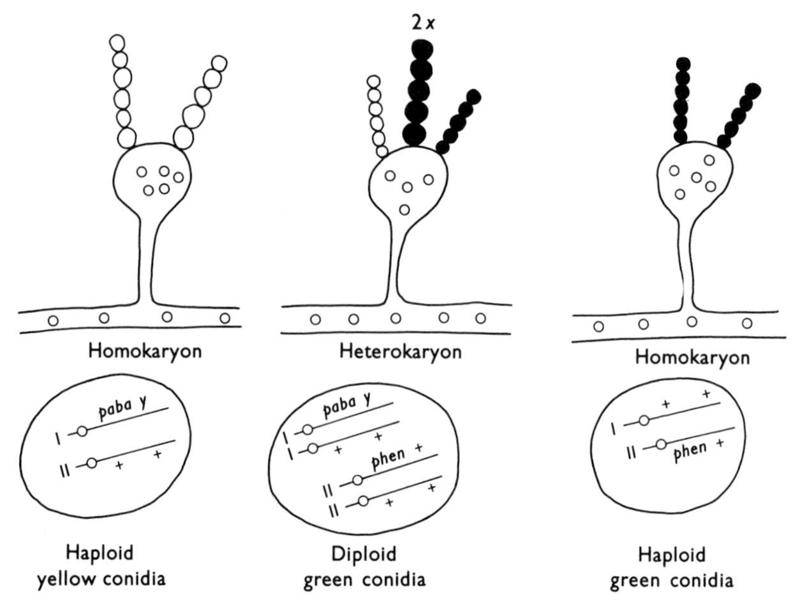

Fig. 7.2 One homokaryon, on chromosome I, has genes *y* (yellow conidia) and *paba* (a nutritional mutant which causes failure of growth in the absence of para-aminobenzoic acid). The other homokaryon carries the mutant *phen* on chromosome II and requires phenylalanine for growth.

or supplemented medium the green mycelium of the diploid breaks up to give yellow sectors. Sub-culturing shows that some of the yellow sectors are haploid but recombinants with respect to the nutritional mutants. Others are diploid and, again, may be recombinants. The haploids are thought to derive from progressive elimination of chromosomes by non-disjunction during mitosis. There is no preference for elimination of the chromosomes of either parent (Pontecorvo and Kafer, 1958).

The diploid yellow sectors (which ultimately tend to become haploid by chromosome elimination) arise by mitotic crossing over (Fig. 7.3). The principle of mapping is much the same as applied in *Drosophila*. With three genes in the one chromosome arm in the order, centromere: $\frac{a\ b\ c}{+\ +\ +}$, only *c* will become homozygous on its own following a cross-over, *b* and *c* the only pair to become homozygous together (double cross-overs are rare). The relative frequencies of the three classes of homozygotes give the map distances between the genes and between the genes and the centromere. These are not strictly comparable with map distances obtained from analysis of sexual ascospores because the total number of diploid nuclei from which the recombinants derive cannot be ascertained. The order of genes as adduced from analysis of mitotic recombination

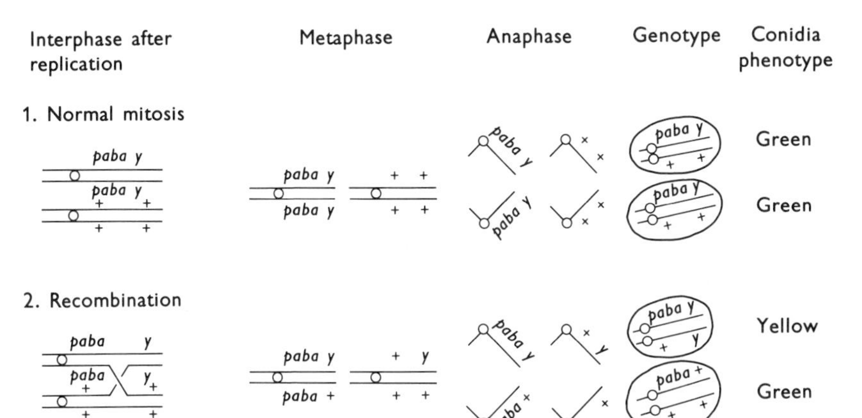

Fig. 7.3 Somatic crossing over in heterozygous diploid *Aspergillus*. The genetic consequences of a normal mitosis are shown in (1). In (2) a homozygous yellow sector results from a single cross-over between the *paba* and the *y* locus.

does, however, correspond with that found from scoring recombinants at meiosis. The relative distances do not correspond, from which it is inferred that the distribution of cross-overs is not the same at mitosis and meiosis.

Together, the formation of the heterokaryon and the nuclear fusion to give diploids, followed by recombination through mitotic crossing over or, indeed, by random chromosome elimination to produce haploid recombinants, constitute the parasexual cycle. The variation generated is probably of particular significance in organisms like the fungi imperfecti which have no sexual cycle. The principles of mapping on the basis of mitotic recombination which were developed in the fungi have proved important when applied to mammalian complements, the human chromosome complement in particular.

7.3 Mouse × man hybrids

The location and mapping of genes by analyzing recombination at meiosis following the standard test crosses are clearly not feasible in *Homo sapiens*. For this reason chromosome maps were until recently based upon painstaking analysis of family pedigrees. The situation was changed dramatically by the development of methods for making intraspecific and interspecific heterokaryons and diploid cell hybrids in mammals.

Heterokaryons and cell hybrids

Harris and Watkins (1965) were the first to produce interspecific heterokaryons in mammalian cells. They contained nuclei from the mouse and man. The hybrid cells arose within a mixture of human *Hela* cells, cultured from cervical carcinoma tissue, and Ehrlich mouse tumour cells which grow as a suspension in the peritoneal cavity of mice. The cell fusions were induced in the presence of a Sendai virus, one of the para-influenza group of myxoviruses. That the multinucleate cells in the mixture were heterokaryotic was verified by labelling the nuclei of the *Hela* cells with tritiated thymidine and showing both labelled and unlabelled nuclei within single cells. Proliferation of the hybrid cells depends on nuclear fusion in binucleate cells (Fig. 7.4). The two nuclei enter mitosis together, a single spindle forms and the daughter cells contain a single nucleus composed of the chromosome complements of both mouse and man. More recently cell hybrids have been produced using normal human cells from a variety of different organs, liver and kidney, also blood cells and nerve cells. In the mouse × man hybrids the fused nuclei initially contain the full diploid chromosome complements of both the parental species. They are, however, unstable much as the diploid nuclei in *Aspergillus* heterokaryons.

The mouse × man cells with hybrid nuclei may be screened out by a method closely allied to that used for screening heterokaryons in fungi where each nucleus compensates for a deficiency in the other. For example, a mixture of human and mouse cells is grown on a medium containing hypoxanthine, aminopterin and thymine. The human cells

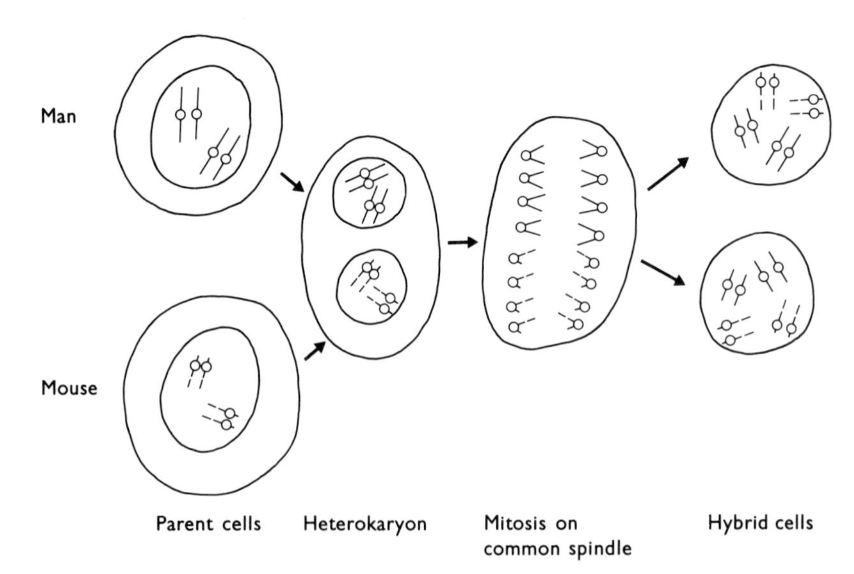

Fig. 7.4 Mouse × man hybrid cell formation. Only two pairs of chromosomes are shown for each parental cell line. Cell fusion leads initially to some binucleate heterokaryons. When both the nuclei enter mitosis synchronously the chromosomes are all attached to a common spindle. Following mitosis both the chromosome sets are contained within the one hybrid cell nucleus.

are blood cells or cells from other non-dividing tissues. The mouse cells are from a strain lacking the enzymes HGPRT (hypoxanthine guanine phosphoribosyltransferase) and TK (thymine kinase). The aminopterin in the medium blocks the synthesis of DNA nucleotides from sugars and amino acids by the regular pathway. The human nuclei which can produce the enzymes HGRRT and TK can synthesize DNA nucleotides by an alternative 'salvage' pathway. Even so the human cells are unable to proliferate because they are from a non-dividing tissue. Neither can the mouse cells multiply because their nuclei, lacking HGPRT and TK, cannot synthesize DNA nucleotides. The fusion nucleus, on the other hand, has the capacity for synthesizing DNA nucleotides, deriving from the human chromosomes, and the capacity for nuclear division, deriving from the mouse chromosomes. For these reasons clones of cells with hybrid nuclei are readily isolated.

Chromosome segregation

In mouse × man cell hybrids the chromosomes of the human complement (the segregant set) are unstable and are eliminated from the cell line. The mouse complement (dominant set) is retained complete. In other situations (Table 7.1), as in the Syrian hamster × mouse hybrids, the dominance relationship may be reversed and it is the mouse set which undergoes elimination.

Table 7.1 Patterns of chromosome segregation in some interspecific cell hybrids

Dominant set		Segregant set
Mouse	×	Man
Chinese hamster	×	Man
Syrian hamster	×	Mouse
Chinese hamster	×	Kangaroo rat
Mouse	×	Rat
Mouse	×	Drosophila

Chromosome elimination of a comparable kind is also found following normal interspecific hybridization between plant species. For example Davies (1958) and Kasha and Kao (1970) found that when they made reciprocal hybrids between a 28 chromosome autotetraploid cultivated barley (*Hordeum vulgare*) and a tetraploid *H. bulbosum*, nearly all the progeny were diploid (i.e. containing gametic chromosome numbers) and resembled the cultivated *H. vulgare*. When crosses were made using the diploid *H. vulgare* the cultured embryos were haploids with chromosomes derived exclusively from the *H. vulgare* maternal parent. The haploids arise by loss of *H. bulbosum* chromosomes during the early mitotic divisions of the embryo. In another cross, using the Chinese

Spring variety of hexaploid cultivated wheat ($2n = 6x = 42$) and both diploid and tetraploid forms of *H. bulbosum* as pollen parents, Barclay (1975) has obtained haploid wheat with 21 chromosomes, again after embryo culture. Chromosome counts at mitotic metaphase in the zygote confirm that fertilization was quite normal and the segregation and loss of *H. bulbosum* chromosomes occurred at the subsequent divisions of the embryo. What determines that the chromosomes of a particular species be eliminated is not known. The same applies to the mammalian cell hybrids (Table 7.1).

Chromosome segregation forms the basis of the mapping procedure for the assignment of genes to chromosomes in the mouse × man hybrids. There is some suggestion that chromosome segregation takes place very soon after cell fusion, possibly at the first mitosis following heterokaryon formation. One of the consequences of this is that the numbers and kinds of chromosomes are fairly constant within young clones, but there is variation between the different clones (Ruddle, 1972). Each of the human chromosomes ($2n = 2x = 46$) is identifiable by its banding pattern (Fig. 1.5, p. 12) and readily distinguished from the chromosomes of the mouse ($2n = 2x = 40$).

Mapping

There is transcription and translation of information from many genes within both man and mouse chromosomes. Where the end product (enzyme) of a gene in the human complement is identifiable, the way to find out which chromosome carries the gene is to examine the chromosomes in clones which do not produce the enzyme and determine which particular chromosome lost by segregation is common to each. The practical aspects have been presented by Ruddle and Kucherlapati (1974) and are summarized in Fig. 7.5.

This kind of genetic analysis is limited to those genes which are expressed in the artificial environment of cell culture and it is clearly not possible to study character differences involving organs let alone the form and functioning of the organisms as a whole. The genes most amenable to analysis are those coding for constitutive enzymes continuously produced throughout the life of most or all cell types. Not surprisingly many of these genes are concerned with basic cellular functions like respiration. Genes specifying inducible enzymes which are more likely to be expressed in specialized differentiated cells can often be investigated if the appropriate cell types concerned, such as liver, kidney or blood cells, are used in the initial cell fusions. Alternatively some genes with highly specialized activities in human cells can, even in cells where they are not normally active, be derepressed by the presence of homologous genes in the mouse complement to permit a linkage analysis.

It is not yet possible to map the positions of genes relative to one another within human chromosomes by somatic crossing over, as it is in some fungi. The approximate positions of some genes, however, in relation to

		Hybrid clones				
		A	B	C	D	E
Human enzymes	I	+	−	−	+	−
	II	−	+	−	+	−
	III	+	−	−	+	−
	IV	+	+	+	−	−
Human chromosomes	1	−	+	−	+	−
	2	+	−	−	+	−
	3	−	−	−	+	+

Fig. 7.5 Experimental procedure for linkage analysis. The analysis is based on the correspondence between the presence or absence of particular human enzymes and chromosomes over a number of hybrid clones. In this hypothetical scheme enzymes I and III are both seen to be linked and assignable to chromosome 2. Enzyme II is assigned to chromosome I. [From Ruddle, F. H. and Kucherlapati, R. S. (1974) *Scientific American*, **231**, 36–44]

chromosome arms or parts of arms can sometimes be ascertained cytologically by the use of translocations, virus-induced lesions or deletions.

8

Conclusion

We have alluded to the variety in the form and behaviour of chromosomes which accompanied the divergence and evolution of species and to the widespread chromosome polymorphism which contributes to adaption within species. Recent evidence for very large scale quantitative variation in the chromosomal DNA, by segmental amplification or deletion, adds a new dimension to the pattern of adaptive chromosome change. The genetic significance of such variation is unresolved. One of the features of surveys describing the quantitative DNA variation among groups of plants and animals is worth emphasizing. It is that quantitative DNA variation *among* species, even closely related species, is commonplace and often on a very large scale, whereas variation *within* species, is uncommon and even then of a very restricted degree. We must conclude that within that group we call a species there are rigid limits to certain kinds of change which in turn argues for a finely adjusted balance among the various chromosome components. It follows that components such as those made up of very highly repetitive DNA segments, seemingly untranscribed and untranscribable, are redundant only in the context of transcription, not in the sense of fitness. Taking a more familiar point, we have known for years that the chromosome number within species is usually constant and so also, is the relative distribution of genetic material and of genes among the chromosome complement. Yet it is by no means clear why this should be so, at least in species comprising widely scattered populations. The point is perhaps best made by dwelling on an exception. In the dog whelk, *Nucella lapillus* (Bantock and Cockayne, 1975), the basic chromosome number and thereby the number of linkage groups varies from one locality to the other, from $2n=26$ to $2n=36$. Another example comes from a gerbil, *Gerbillus pyramidum* (Wahrman and Gourevitz, 1972). The creation or disruption of linkages is on the face of it a powerful adaptive option. Yet the option is rarely taken up. We must assume that in general the architecture of each chromosome has evolved to accommodate its particular role and activity. We have little understanding of how this is achieved but the rigidity with regard to chromosome number and DNA content may well be related. The amplified DNA segments could well serve as the 'bricks and mortar' which determine the construction and thereby the functioning of the chromosome. While large scale changes in chromosome architecture are restricted within a species it is manifestly clear that speciation within many groups of plants and animals is accompanied or succeeded by changes of cataclysmic proportions. To say the least there is a surprising *discontinuity* of chromosome organization *among* species.

Finally, it is worth considering the implications of recent information about the molecular organization of the chromosomes to the formal concepts of gene action and interaction. If, as is suggested above, the untranscribed fraction of the chromosomal DNA is nevertheless instrumental in regulating by one means or another the activity of 'structural' genes of major effect or of genes with minor effects which make up a polygenic system, the untranscribed fractions must be considered no less 'factors' of inheritance than Mendel's genes in peas. The distribution of such material among the chromosomes, in terms of linkage, its mode of action and of interaction within the nucleus, will ultimately need to be incorporated into any model which seeks to describe the transmission and expression of genetic information.

References

AHLOOWALIA, B. S. (1971). Frequency, origin and survival of aneuploids in tetraploid ryegrass. *Genetica*, **42**, 129–38.

ASHBURNER, M. (1969). Genetic control of puffing in polytene chromosomes. *Chromosomes Today*, **2**, 99–106.

ASAO, T. (1969). Behaviour of histones and cytoplasmic basic proteins during embryogenesis of the Japanese newt, *Triturus pyrrhogaster*. *Exp. Cell Res.*, **58**, 243–52.

AVERY, O. T., MACLEOD, C. M. and MCCARTHY, M. (1944). Studies on the chemical nature of the substance inducing transformation of pneumococcal types. *J. exp. Med.*, **79**, 137–58.

AYONOADU U. W. U. and REES H. (1968). The influence of *B* chromosomes on chiasma frequencies in Black Mexican Sweet Corn. *Genetica*, **39**, 75–81.

BACHMANN K., GOIN, O. B. and GOIN, C. J. (1972). Nuclear DNA amounts in vertebrates. *Brookhaven Symp. Biol.*, **23**, 419–50.

BANTOCK, C. R. and COCKAYNE, W. C. (1975). Chromosome polymorphism in *Nucella lapillus*. *Heredity*, **34**, 231–46.

BARCLAY, I. R. (1975). High frequencies of haploid production in wheat (*Triticum aestivum*) by chromosome elimination. *Nature*, **256**, 410–11.

BARLOW, B. A. and WIENS, D. (1975). Permanent translocation heterozygosity in *Viscum hildebrandtii* Engl. and *V. engleri* Tiegh. (*Viscaceae*) in East Africa. *Chromosoma*, **53**, 265–72.

BEADLE, G. W. (1930). Genetical and cytological studies of mendelian asynapsis in *Zea mays*. *Cornell Univ. Agric. Exp. Stat. Mem.*, **129**, 1–23.

BELOZERSKY, A. N. and SPIRIN, A. S. (1960). In *The Nucleic Acids*, Vol. 3, ed. CHARGAFF, E. and DAVIDSON, J. N., Academic Press, Inc., New York and London.

BENNETT, M. D. (1971). The duration of meiosis. *Proc. R. Soc. B.*, **178**, 277–99.

BENNETT, M. D. (1972). Nuclear DNA content and minimum generation time in herbaceous plants. *Proc. R. Soc. B.*, **181**, 109–35.

BENNETT, M. D. and REES, H. (1969). Induced and developmental variation in chromosomes of meristematic cells. *Chromosoma*, **27**, 226–44.

BISHOP, J. O. and FREEMAN, K. B. (1973). DNA sequences neighboring the duck hemoglobin genes. *Cold Spring Harb. Symp. quant. Biol.*, **38**, 707–16.

BLUMENTHAL, A. B., KRIEGSTEIN, H. J. and HOGNESS, D. S. (1973). The units of DNA replication in *Drosophila melanogaster* chromosomes. *Cold Spring Harb. Symp. quant. Biol.*, **38**, 205–23.

BRIDGES, C. B. (1922). The origin of variations in sexual and sex-limited characters. *Am. Nat.*, **56**, 51–63.

BRIDGES, C. B. (1936). The bar 'gene' a duplication. *Science, N.Y.*, **83**, 210–11.

BRITTEN, R. J. and KOHNE, D. E. (1968). Repeated sequences in DNA. *Science, N.Y.*, **161**, 529–40.

BROWN, S. W. and ZOHARY, D. (1955). The relationship of chiasmata and crossing over in *Lilium formosanum*. *Genetics, Princeton*, **40**, 850–73.

CALLAN, H. G. (1972). Replication of DNA in the chromosomes of eukaryotes. *Proc. R. Soc. B.*, **181**, 19–41.

CALLAN, H. G. and LLOYD, L. (1960). Lampbrush chromosomes of crested newts *Triturus cristatus* (Laurenti). *Phil. Trans. R. Soc. B.*, **243**, 135–219.

CAMERON, F. M. and REES, H. (1967). The influence of *B* chromosomes on meiosis in *Lolium*. *Heredity*, **22**, 446–50.

CATCHESIDE, D. G. (1977). *The Genetics of Recombination*. Edward Arnold, London.

CHUKSANOVA, N. (1939). Karyotypes of pollen grains in triploid *Crepis capillaris*. *Comp. Rends. (Doklady) Acad. Sci. U.S.S.R.*, **25**, 232–5.

CRANE, M. B. and LAWRENCE, W. J. C. (1934). *The Genetics of Garden Plants*. Macmillan, London.

CROWLEY, J. G. and REES, H. (1968). Fertility and selection in tetraploid *Lolium*. *Chromosoma*, **24**, 300–8.

DANIELLI, J. F. (1958). Studies of inheritance in amoebae by the technique of nuclear transfer. *Proc. R. Soc., B.*, **148**, 321–31.

DARLINGTON, C. D. (1935). The internal mechanics of the chromosomes II. Prophase pairing at meiosis in *Fritillaria*. *Proc. R. Soc. B.*, **118**, 59–73.

DARLINGTON, C. D. (1937). *Recent Advances in Cytology*. J. & A. Churchill, London.

DARLINGTON, C. D. (1956). *Chromosome Botany*. George Allen and Unwin, London.

DARLINGTON, C. D. (1958). *Evolution of Genetic Systems*, 2nd ed. Oliver & Boyd, Edinburgh.

DARLINGTON, C. D., HAIR, J. B. and HURCOMBE, R. (1951). The history of the garden Hyacinths. *Heredity*, **5**, 233–52.

DARLINGTON, C. D. and LA COUR, L. F. (1950). Hybridity selection in *Campanula*. *Heredity*, **4**, 217–48.

DARLINGTON, C. D. and MATHER, K. (1949). *The Elements of Genetics*. Allen and Unwin, London.

DAVIDSON, E. H., GRAHAM, D. E., NEUFIELD, B. R., CHAMBERLIN, M. E., AMENSON, C. S., HOUGH, B. R. and BRITTEN, R. J. (1973). Arrangements and characterisation of repetitive sequence elements in animal DNAs. *Cold Spring Harb. Symp. quant. Biol.*, **38**, 295–301.

DAVIES, D. R. (1958). Male parthenogenesis in barley. *Heredity*, **12**, 493–8.

DHIR, N. K. and MIKSCHE, J. P. (1974). Intraspecific variation of nuclear DNA content in *Pinus resinosa*. Ait. *Can. J. Genet. Cytol.*, **16**, 77–83.

DOBZHANSKY, T. (1943). Genetics of natural populations IX. Temporal changes in the composition of populations of *Drosophila pseudoobscura*. *Genetics, Princeton*, **28**, 162–86.

DOBZHANSKY T. (1948). Genetics of natural populations. XVIII. Experiments on chromosomes of *Drosophila pseudoobscura* from different geographic regions. *Genetics, Princeton*, **33**, 588–602.

DODDS, K. S. (1965). The history and relationships of cultivated potatoes, 123–41. In *Essays on Crop Plant Evolution*, ed. J. Hutchinson, Cambridge University Press.

DOVER, G. A. and HENDERSON, S. A. (1975). No detectable satellite DNA in supernumerary chromosomes of the grasshopper *Myrmeleotettix*. *Nature*, **259**, 57–9.

DOVER, G. A. and RILEY, R. (1972). Prevention of pairing of homoeologous meiotic chromosomes of wheat by an activity of supernumerary chromosomes of *Aegilops*. *Nature*, **240**, 159–61.

DUPRAW, E. J. (1970). *DNA and chromosomes*. Holt, Rhinehart and Winston, Inc., New York.

EVANS, G. M. and MACEFIELD, A. J. (1972). Suppression of homoeologous pairing by *B*-chromosomes in a *Lolium* species hybrid. *Nature New Biol.*, **236**, 110–11.

EVANS, G. M., DURRANT, A. and REES, H. (1966). Associated nuclear changes in the induction of Flax Genotrophs. *Nature*, **212**, 697–9.

EVANS, G. M., REES, H., SNELL, C. and SUN, S. (1972). The relationship between nuclear DNA amount and the duration of the mitotic cycle. *Chromosomes Today*, **3**, 24–31.

EVANS, H. J. and SUMNER, A. T. (1973). Chromosome architecture; morphological and molecular aspects of longitudinal differentiation. *Chromosomes Today*, **4**, 15–33.

FINCHAM, J. R. S. and DAY, P. R. (1971). *Fungal Genetics*. Blackwell, Oxford.

FLAVELL, R. B. and RIMPAU, J. (1975). Ribosomal RNA genes and supernumerary *B*-chromosomes of rye. *Heredity*, **35**, 127–31.

FLAVELL, R. B. and SMITH, D. B. (1974). Variation in nucleolar organiser rRNA gene multiplicity in wheat and rye. *Chromosoma*, **47**, 327–34.

FOGEL, S. and MORTIMER, R. K. (1969). Informational transfer in meiotic gene conversion. *Proc. Natn. Acad. Sci. U.S.A.*, **62**, 96–103.

FOX, D. P. (1971). The replicative status of heterochromatic and euchromatic DNA in two somatic tissues of *Dermestes maculatus* (*Dermestidae: Coleoptera*). *Chromosoma*, **33**, 183–95.

FRÖST, S. (1958). Studies of the genetical effects of accessory chromosomes in *Centaurea scabiosa*. *Hereditas*, **44**, 112–22.

GALL, J. G. (1956). *Brookhaven Symp. Biol.*, **8**, 17–32.

GALL, J. G. (1968). Differential synthesis of the genes for ribosomal RNA during amphibian oogenesis. *Proc. Natn. Acad. Sci. U.S.A.*, **60**, 553–60.

GALL, J. G. and CALLAN, H. G. (1962). H^3 uridine incorporation in lampbrush chromosomes. *Proc. Natn. Acad. Sci. U.S.A.*, **48**, 562–70.

GALL, J. G., COHEN, E. H. and ATHERTON, D. D. (1973). The satellite DNAs of *Drosophila virilis*. *Cold Spring Harb. Symp. quant. Biol.*, **38**, 417–21.

GERSTEL, D. U. and BURNS, J. A. (1966). Chromosomes of unusual length in hybrids between two species of *Nicotiana*. *Chromosomes Today*, **1**, 41–56.

GRANT, V. (1959). The regulation of recombination in plants. *Cold Spring Harb. Symp. quant. Biol.*, **23**, 337–63.

GRELL, R. F. and CHANDLEY, A. C. (1965). Evidence bearing on the coincidence of exchange and DNA replication in the oocyte of *Drosophila melanogaster*. *Proc. Natl. Acad. Sci. U.S.A.*, **53**, 1340–6.

GRIFFITH, F. (1928). Significance of pneumococcal types. *J. Hyg., Camb.*, **27**, 113–59.

GURDON, J. B. (1973). Nuclear transplantation and regulation of cell processes. *Br. med. Bull.*, **29**, 259–62.

HAMERTON, J. L. (1971). *Human Cytogenics*, Vol. I. Academic Press, Inc., New York and London.

HARINARAYANA, G. and MURTY, B. R. (1971). Cytological regulation of recombination in *Pennisetum* and *Brassica*. *Cytologia*, **36**, 435–48.

HARRIS, H. and WATKINS, J. F. (1965). Hybrid cells derived from mouse and man: artificial heterokaryons of mammalian cells from different species. *Nature*, **205**, 640–6.

HAYES, W. (1968). *The Genetics of Bacteria and their Viruses*, 2nd ed. Blackwell, Oxford.

HAZARIKA, M. H. and REES, H. (1967). Genotypic control of chromosomes behaviour in rye X. Chromosome pairing and fertility in autotetraploids. *Heredity*, **22**, 317–32.

HENDERSON, S. A. (1963). Chiasma distribution at diplotene in a locust. *Heredity*, **18**, 173–90.

HENDERSON, S. A. (1970). The time and place of meiotic crossing-over. *A. Rev. Genet.*, **4**, 295–324.

HERSHEY, A. D. (1955). An upper limit to the protein content of the germinal substance of bacteriophage T2. *Virology*, **1**, 108–27.

HERSHEY, A. D. and CHASE, M. C. (1952). Independent functions of viral protein and nucleic acid in growth of bacteriophage. *J. gen. Physiol.*, **36**, 39–56.

HEWITT, G. M. (1973). The integration of supernumerary chromosomes into the Orthopteran genome. *Cold Spring Harb. Symp. quant. Biol.*, **38**, 183–94.
HEWITT, G. M. and JOHN, B. (1972). Inter-population sex chromosome polymorphism in the grasshopper *Podisma pedestris* II. Population parameters. *Chromosoma*, **37**, 23–42.
HOLLIDAY, R. (1964). A mechanism for gene conversion in fungi. *Genet. Res.*, **5**, 282–304.
HOWELL, S. H. and STERN, H. (1971). The appearance of DNA breakage and repair activities in the synchronous meiotic cycle of *Lilium*. *J. molec. Biol.*, **55**, 357–78.
HUTTO, J. (1974). The response of diploid and tetraploid rye genotypes to phosphate treatments and to cold temperature. *J. agric. Sci., Camb.*, **82**, 353–6.
HUTTON, J. R. and THOMAS, C. A. (1975). The origin of folded DNA rings from *Drosophila melanogaster*. *J. molec. Biol.*, **98**, 425–38.
JOHN, B. (1973). The cytogenetic systems of grasshoppers and locusts II. The origin and evolution of supernumerary segments. *Chromosoma*, **44**, 123–46.
JOHN, B. and HEWITT, G. M. (1965). The *B*-chromosome system of *Myrmeleotettix maculatus* (Thunb.) I. The mechanics. *Chromosoma*, **16**, 548–78.
JOHN, B. and HEWITT, G. M. (1968). Patterns and pathways of chromosome evolution within the *Orthoptera*. *Chromosoma*, **25**, 40–74.
JOHN, B. and LEWIS, K. R. (1957). Studies on *Periplaneta americana*. II. Interchange heterozygosity in isolated populations. *Heredity*, **11**, 11–22.
JOHN, B. and LEWIS, K. R. (1965). *The Meiotic System, Vol. VI F1, Protoplasmatologia*. Springer-Verlag, Vienna.
JONES, G. H. (1967). The control of chiasma distribution in rye. *Chromosoma*, **22**, 69–90.
JONES, G. H. (1971). The analysis of exchanges in tritium-labelled meiotic chromosomes II. *Stethophyma grossum*. *Chromosoma*, **34**, 367–82.
JONES, K. (1974). Chromosome evolution by Robertsonian translocation in *Gibasis* (*Commelinaceae*). *Chromosoma*, **45**, 353–68.
JONES, K. W. (1970). Chromosomal and nuclear location of mouse satellite DNA in individual cells. *Nature*, **225**, 912–15.
JONES, K. W. and ROBERTSON, F. W. (1970). Localisation of reiterated nucleotide sequences in *Drosophila* and mouse by *in situ* hybridisation of complementary RNA. *Chromosoma*, **31**, 331–45.
JONES, R. N. (1975). *B*-chromosome systems in flowering plants and animal species. *Int. Rev. Cytol.*, **40**, 1–100.
JONES, R. N. and BROWN, L. M. (1976). Chromosome evolution and DNA variation in *Crepis*. *Heredity*, **36**, 91–104.
JONES, R. N. and REES, H. (1968). Nuclear DNA variation in *Allium*. *Heredity*, **23**, 591–605.
JONES, R. N. and REES, H. (1969). An anomalous variation due to *B*-chromosomes in rye. *Heredity*, **24**, 265–71.
JUDD, B. H. and YOUNG, M. W. (1973). An examination of the one cistron:one chromomere concept. *Cold Spring Harb. Symp. quant. Biol.*, **38**, 573–9.
KARPECHENKO, G. D. (1928). Polyploid hybrids of *Raphanus sativa* L. × *Brassica oleracea* L. *Zeitschr. ind Abstamm. Vererbungsl.*, **39**, 1–7.
KASHA, K. J. and KAO, K. N. (1970). High frequency haploid production in barley (*Hordeum valgare* L.). *Nature*, **225**, 874–5.
KAVENOFF, R., KLOTZ, L. C. and ZIMM, B. H. (1973). On the nature of chromosome-sized DNA molecules. *Cold Spring Harb. Symp. quant. Biol.*, **38**, 1–8.
KAYANO, H. (1971). Accumulation of *B*-chromosomes in the germ line of *Locusta migratoria*. *Heredity*, **27**, 119–23.
KEYL, H. G. (1965). Duplikationen von unterreinheiten der chromosomalen DNA während der evolution von *Chironomous thummi*. *Chromosoma*, **17**, 139–80.

KIRK, D. and JONES, R. N. (1970). Nuclear genetic activity in *B*-chromosome rye in terms of the quantitative interrelations between nuclear protein, nuclear RNA and histone. *Chromosoma*, **31**, 241–54.

KIRK, J. T. O., REES, H. and EVANS, G. M. (1970). Base composition of nuclear DNA within the genus *Allium*. *Heredity*, **25**, 507–12.

KITANI, Y., OLIVE, L. S. and EL-ANI, A. S. (1962). Genetics of *Sordaria fimicola* V. Aberrant segregation at the *g* locus. *Am. J. Bot.*, **49**, 697–706.

KORNBERG, R. D. and THOMAS, J. O. (1974). Chromatin structure: oligomeres of the histones. *Science, N.Y.*, **184**, 865–8.

LAWRENCE, C. W. (1961). The effect of radiation on chiasma formation in *Tradescantia*. *Radiation Bot.*, **1**, 92–6.

LEWIN, B. (1975). The nucleosome: subunit of mammalian chromatin. *Nature*, **254**, 651–3.

LEWIS, K. R. (1967). Polyploidy and plant improvement. *The Nucleus*, **10**, 99–110.

LEWIS, K. R. and JOHN, B. (1963). *Chromosome Markers*. J. and A. Churchill, Ltd., London.

LEWIS, K. R. and JOHN, B. (1970). *The Organization of Heredity*. Edward Arnold, London.

LIMA-DE-FARIA, A. (1954). Chromosome gradient and chromosome field in *Agapanthus*. *Chromosoma*, **6**, 330–70.

LIMA-DE-FARIA, A. (1959). Differential uptake of tritiated thymidine into hetero- and euchromatin in *Melanoplus* and *Secale*. *J. Biophys. Biochem. Cytol.*, **6**, 457–66.

LISSOUBA, P., MOUSSEAU, J., RIZET, G. and ROSSIGNOL, J. L. (1962). Fine structure of genes in the ascomycete *Ascobolus immersus*. *Adv. Genet.*, **11**, 343–80.

LOENING, U. E., JONES, K. W. and BIRNSTIEL, M. L. (1969). Properties of the ribosomal RNA precursor in *Xenopus laevis*; comparison to the precursor in mammals and in plants. *J. molec. Biol.*, **45**, 353–66.

LONGLEY, A. E. (1927). Supernumerary chromosomes in *Zea mays*. *J. Agric. Res.*, **35**, 769–84.

LYON, M. F. (1963). Attempts to test the inactive-X theory of dosage compensation in mammals. *Genet. Res.*, **4**, 93–103.

MCPHEE, C. P. and ROBERTSON, A. (1970). The effect of suppressing crossing-over on the response to selection in *Drosophila melanogaster*. *Genet. Res.*, **16**, 1–16.

MARTIN, P. G. (1966). Variation in the amounts of nucleic acids in the cells of different species of higher plants. *Expl. Cell. Res.*, **44**, 84–94.

MATHER, K. (1936). Segregation and linkage in autotetraploids. *J. Genet.*, **31**, 287–314.

MATHER, K. (1937). The determination of position in crossing over. II. The chromosome length chiasma frequency relation. *Cytologia* (Tokyo) Fujii Jub. vol., 514–26.

MATHER, K. (1938). Crossing over. *Biol. Rev.*, **13**, 252–92.

MATHER, K. (1973). *Genetical Structure of Populations*. Chapman and Hall, London.

MATSUDA, T. (1970). On the accessory chromosomes of *Aster*. I. The accessory chromosomes of *Aster ageratoides* group. *J. Sci. Hiroshima Univ.*, **13**, 1–63.

MIRSKY, A. E. and OSAWA, S. (1961). *The Cell*, Vol. II, Academic Press.

MILLER, O. L. (1965). Fine structure of lampbrush chromosomes. *Natn. Cancer Inst. Monograph*, **18**, 79–89.

MITTWOCH, U. (1967). *Sex chromosomes*. Academic Press, Inc., New York.

MOENS, P. B. (1969a). The fine structure of meiotic chromosome polarization and pairing in *Locusta migratoria* spermatocytes. *Chromosoma*, **28**, 1–25.

MOENS, P. B. (1969b). The fine structure of meiotic chromosome pairing in the triploid, *Lilium tigrinum*. *J. Cell Biol.*, **40**, 273–9.

MORESCALCHI, A. (1973). Amphibia. *In Cytotaxonomy and Vertebrate Evolution*, ed. A. B. Chiarelli and E. Capanna, Academic Press, Inc., New York and London, pp. 233–348.

MORRIS, T. (1968). The *XO* and *OY* chromosome constitutions in the mouse. *Genet. Res.*, **12**, 125–37.
MOSES, M. J. (1956). Chromosomal structures in crayfish spermatocytes. *J. Biophys. Biochem. Cytol.*, **2**, 215–17.
MOSS, J. P. (1966). The adaptive significance of *B*-chromosomes in rye. *Chromosomes Today*, **1**, 15–23.
MULLER, H. J. and KAPLAN, W. D. (1966). The dosage compensation of *Drosophila* and mammals as showing the accuracy of the normal type. *Genet. Res.*, **8**, 41–59.
MÜNTZING, A. (1963). Effects of accessory chromosomes in diploid and tetraploid rye. *Hereditas*, **49**, 371–426.
NARAYAN, R. K. J. and REES, H. (1976). Nuclear DNA variation in *Lathyrus*. *Chromosoma*, **54**, 141–54.
OHNO, S. (1969). Evolution of sex chromosomes in mammals. *Ann. Rev. Genetics*, **3**, 495–524.
OHNO, S. (1970). *Evolution by Gene Duplication*. Allen and Unwin, London.
OLINS, A. L. and OLINS, D. E. (1974). Spheroid chromatin units (v bodies). *Science, N.Y.*, **183**, 330–2.
OUDET, P., GROSS-BELLARD, M. and CHAMBON, P. (1975). Electron microscopic and biochemical evidence that chromatin structure is a repeating unit. *Cell*, **4**, 281–300.
PARDUE, M. L. and GALL, J. G. (1970). Chromosomal localisation of mouse satellite DNA. *Science, N.Y.*, **168**, 1356–8.
PARKER, J. S. (1975). Chromosome-specific control of chiasma formation. *Chromosoma*, **49**, 391–406.
PEACOCK, W. J. (1970). Replication, recombination, and chiasmata in *Goniaea australasiae* (Orthoptera: Acrididae). *Genetics*, **65**, 593–617.
PEACOCK, W. J., BRUTLAG, D., GOLDRING, E., APPELS, R., HINTON, C. W. and LINDSLEY, D. L. (1973). The organisation of highly repeated DNA sequences in *Drosophila melanogaster* chromosomes. *Cold Spring Harb. Symp. quant. Biol.*, **38**, 405–16.
PEARSON, G. G., TIMMIS, J. N. and INGLE, J. (1974). The differential replication of DNA during plant development. *Chromosoma*, **45**, 281–94.
PERRY, P. E. and JONES, G. H. (1974). Male and female meiosis in grasshoppers. I. *Stethophyma grossum*. *Chromosoma*, **47**, 227–36.
PONTECORVO, G. (1953). The genetics of *Aspergillus nidulans*. *Adv. Genet.*, **5**, 141–238.
PONTECORVO, G. and KAFER, E. (1958). Genetic analysis based on mitotic recombination. *Adv. Genet.*, **9**, 71–103.
RANDOLPH, L. F. (1928). Types of supernumerary chromosomes in maize. *Anat. Rec.*, **41**, 102.
REDDI, K. K. (1959). The arrangement of purine and pyrimidine nucleotides in tobacco mosaic virus nucleotides. *Proc. Natn. Acad. Sci. U.S.A.*, **45**, 293–300.
REES, H. (1964). The question of polyploidy in the *Salmonidae*. *Chromosoma*, **15**, 275–9.
REES, H. (1972). DNA in higher plants. *Brookhaven Symp. Biol.*, **23**, 394–418.
REES, H., CAMERON, F. M., HAZARIKA, M. H. and JONES, G. H. (1966). Nuclear variation between diploid angiosperms. *Nature*, **211**, 828–30.
REES, H. and DALE, P. J. (1974). Chiasmata and variability in *Lolium* and *Festuca* populations. *Chromosoma*, **47**, 335–51.
REES, H. and HAZARIKA, M. H. (1969). Chromosome evolution in *Lathyrus*. *Chromosomes Today*, **2**, 158–65.
REES, H. and HUTCHINSON, J. (1973). Nuclear DNA variation due to *B*-chromosomes. *Cold Spring Harb. Symp. quant. Biol.*, **38**, 175–82.
REES, H. and JONES, G. H. (1967). Chromosome evolution in *Lolium*. *Heredity*, **22**, 1–18.

REES, H. and JONES, R. N. (1971). Chromosome gain in higher plants, in: *Cellular Organelles and Membranes in Mental Retardation*, ed. P. F. Benson. Churchill Livingstone, Edinburgh and London, pp. 185–208.

REES, H. and JONES, R. N. (1972). The origin of the wide species variation in nuclear DNA content. *Int. Rev. Cytol.*, **32**, 53–92.

REES, H. and SUN, S. (1965). Chiasma frequency and the disjunction of interchange associations in rye. *Chromosoma*, **16**, 500–10.

REES, H. and THOMPSON, J. B. (1956). Genotypic control of chromosome behaviour in rye. III. Chiasma frequency in homozygotes and heterozygotes. *Heredity*, **3**, 409–24.

RHOADES, M. M. (1968). Studies on the cytological basis of crossing over, in: *Replication and Recombination of Genetic Material*, ed. W. J. Peacock and R. D. Brock. Australian Academy of Sciences, Canberra, pp. 229–41.

RHOADES, M. M. and DEMPSEY, E. (1973). Chromatin elimination induced by the *B*-chromosome of maize. *J. Heredity*, **64**, 13–18.

RICK, C. M. and BARTON, D. W. (1954). Cytological and genetic identification of the primary trisomics of the tomato. *Genetics*, **39**, 640–66.

RILEY, R. and CHAPMAN, V. (1958). Genetic control of the cytologically diploid behaviour of hexaploid wheat. *Nature*, **182**, 713–15.

RITOSSA, F. M. and SCALA, G. (1969). Equilibrium variations in the redundancy of rDNA in *Drosophila melanogaster*. *Genetics Suppl., Princeton*, **61**, 305–17.

RITOSSA, F. M. and SPIEGELMAN, S. (1965). Localisation of DNA complementary to ribosomal RNA in the nucleolus organizer region of *Drosophila melanogaster*. *Proc. Natn. Acad. Sci. U.S.A.*, **53**, 737–45.

ROBERTSON, W. R. B. (1916). Chromosome studies. I. Taxonomic relationships shown in the chromosomes of *Tettigidae* and *Acrididae*: V-shaped chromosomes and their significance in *Acrididae*, *Locustidae*, and the *Gryllidae*: chromosomes and variation. *J. Morph.*, **27**, 179–331.

ROMAN, H. (1948). Directed fertilisation in maize. *Proc. natl. Acad. Sci. U.S.A.*, **34**, 36–42.

ROSSEN, J. M. and WESTERGAARD, M. (1966). Studies on the mechanism of crossing-over. II. Meiosis and the time of chromosome replication in the Ascomycete *Neottiella rutilans* (Fr.) Dennis. *Comp. Rend.*, **35**, 233–60.

RUDDLE, F. H. (1972). Linkage analysis using somatic cell hybrids. *Adv. Hum. Genet.*, **3**, 173–235.

RUDDLE, F. H. and KUCHERLAPATI, R. S. (1974). Hybrid cells and human genes. *Scientific American*, **231**, 36–44.

SAGAR, J. and RYAN, F. J. (1963). *Cell Heredity*. John Wiley, Chichester.

SANNOMIYA, M. (1962). Intra-individual variation in number of *A*- and *B*-chromosomes in *Pantanga japonica*. *Chromosome Information Service*, **3**, 30–2.

SCHACHAT, F. H. and HOGNESS, D. S. (1973). Repetitive sequences in isolated Thomas circles from *Drosophila melanogaster*. *Cold Spring Harb. Symp. quant. Biol.*, **38**, 371–81.

SCHINDLER, A. M. and MIKAMO, K. (1970). Triploidy in man. Report of a case and a discussion on etiology. *Cytogenetics*, **9**, 116–30.

SELIGY, V. and MIYAGI, M. (1969). Studies of template activity of chromatin isolated from metabolically active and inactive cells. *Exp. Cell Res.*, **58**, 27–34.

SHARMAN, G. B. and BARBER, H. N. (1952). Multiple sex chromosomes in the marsupial *Potorus*. *Heredity*, **6**, 345–55.

SIMONSEN, Ø. (1973). Cytogenetic investigations in diploid and autotetraploid populations of *Lolium perenne* L. *Hereditas*, **75**, 157–88.

SIMONSEN, Ø. (1975). Cytogenetic investigations in diploid and autotetraploid populations of *Festuca pratensis*. *Hereditas*, **79**, 73–108.

SMITH, D. B. and FLAVELL, R. B. (1975). Characterisation of the wheat genome by renaturation kinetics. *Chromosoma*, **50**, 223–42.

SOUTHERN, D. I. (1970). Polymorphism involving heterochromatic segments in *Metrioptera brachyptera*. *Chromosoma*, **30**, 154–68.
SOUTHERN, E. M. (1970). Base sequence and evolution of guinea-pig satellite DNA. *Nature*, **227**, 794–8.
SOUTHERN, E. M. (1974). Eukaryotic DNA, in: *Biochemistry of Nucleic Acids*, ed. K. Burton. Butterworths, London, pp. 101–39.
SPARROW, A. H., PRICE, H. J. and UNDERBRINK, A. G. (1972). A survey of DNA content per cell and per chromosome in prokaryotic and eukaryotic organisms; some evolutionary considerations. *Brookhaven Symp. Biol.*, **23**, 451–94.
STEBBINS, G. L. (1950). *Variation and Evolution in Plants*. Columbia University Press.
STEBBINS, G. L. (1966). Chromosomal variation and evolution. *Science, N.Y.*, **152**, 1463–9.
STERN, C. (1936). Somatic crossing over and segregation in *Drosophila melanogaster*. *Genetics*, **21**, 625–730.
STERN, C. (1970). Somatic recombination within the white locus of *Drosophila melanogaster*. *Genetics*, **62**, 573–81.
STERN, H. and HOTTA, Y. (1973). Biochemical controls of meiosis. *Ann. Rev. Genet.*, **7**, 37–66.
STONE, W. S. (1955). Genetic and chromosomal variability in *Drosophila*. *Cold Spring Harb. Symp. quant. Biol.*, **20**, 256–70.
SUEOKA, N. (1961). Variation and heterogeneity of base composition of deoxyribonucleic acids: a compilation of old and new data. *J. molec. Biol.*, **3**, 31–40.
SUN, S. and REES, H. (1964). Genotypic control of chromosome behaviour in rye VII. Unadaptive heterozygotes. *Heredity*, **19**, 357–67.
SWANSON, C. P. (1960). *Cytology and Cytogenetics*. Macmillan, London.
TAYLOR, J. H., WOODS, P. S. and HUGHES, W. L. (1957). The organisation and duplication of chromosomes as revealed by autoradiographic studies using tritium-labelled thymidine. *Proc. Natn. Acad. Sci. U.S.A.*, **43**, 122–8.
THOMAS, C. A. (1970). The theory of the master gene, in: *The Neurosciences: Second Study Program*, ed. F. O. Schmitt. Rockefeller University Press, New York, pp. 973–98.
THOMAS, C. A., PYERITZ, R. E., WILSON, D. A., DANCIS, B. M., LEE, C. S., BICK, M. D., HUANG, H. L. and ZIMM, B. H. (1973). Cyclodromes and palindromes in chromosomes. *Cold Spring Harb. Symp. quant. Biol.*, **38**, 353–70.
THOMPSON, J. B. (1956). Genotypic control of chromosome behaviour in rye. II. Disjunction at meiosis in interchange heterozygotes. *Heredity*, **10**, 99–108.
TIMMIS, J. N., SINCLAIR, J. and INGLE, J. (1972). Ribosomal-RNA genes in euploids and aneuploids of Hyacinth. *Cell Differentiation*, **1**, 335–9.
TIMMIS, J. N. and REES, H. (1971). A pairing restriction at pachytene upon multivalent formation in autotetraploids. *Heredity*, **26**, 269–75.
TIMMIS, J. N. and INGLE, J. (1973). Environmentally induced changes in rDNA gene redundancy. *Nature New Biol.*, **244**, 235–6.
TOBGY, H. A. (1943). A cytological study of *Crepis fuliginosa*, *C. neglecta* and their F_1 hybrid and its bearing on the mechanism of phylogenic reduction in chromosome number. *J. Genet.*, **45**, 67–111.
VOSA, C. G. and BARLOW, P. W. (1972). Meiosis and *B*-chromosomes in *Listera ovata* (Orchidaceae), *Caryologia*, **25**, 1–8.
WAHRMAN, J. and GOUREVITZ, P. (1971). Extreme chromosomal variability in a colonising rodent. *Chromosomes Today*, **4**, 399–424.
WALTERS, J. L. (1942). Distribution of structural hybrids in *Paeonia californica*. *Am. J. Bot.*, **29**, 270–5.
WARD, E. J. (1973). Nondisjunction: localisation of the controlling site in the maize *B*-chromosome. *Genetics, Princeton*, **73**, 387–91.
WASSERMAN, M. (1963). Cytology and phylogeny of *Drosophila*. *Am. Nat.*, **97**, 333–352.

WESTERGARRD, M. and VON WETTSTEIN, D. (1972). The synaptonemal complex. *Ann. Rev. Genet.*, **6**, 71–110.

WHITE, M. J. D. (1956). Adaptive chromosomal polymorphism in an Australian grasshopper. *Evolution*, **10**, 298–313.

WHITE, M. J. D. (1957a). Cytogenetics of the grasshopper *Moraba scurra*, I. Meiosis of interracial and interpopulation hybrids. *Aust. J. Zool.*, **5**, 285–304.

WHITE, M. J. D. (1957b). Cytogenetics of the grasshopper *Moraba scurra*. II. Heterotic systems and their interaction. *Aust. J. Zool.*, **5**, 305–37.

WHITE, M. J. D. (1957c). Cytogenetics of the grasshopper *Moraba scurra*. IV. Heterozygosity for 'elastic constrictions'. *Aust. J. Zool.*, **5**, 348–54.

WHITE, M. J. D. (1973). *Animal Cytology and Evolution*. Cambridge University Press.

WHITEHOUSE, H. L. K. (1963). A theory of crossing-over by means of hybrid deoxyribonucleic acid. *Nature*, **199**, 1034–40.

WHITEHOUSE, H. L. K. (1969). *Towards an Understanding of the Mechanism of Heredity*, 2nd ed. Edward Arnold, London.

WIMBER, D. E. and STEFFENSEN, D. M. (1973). Localisation of gene function. *Ann. Rev. Gent.*, **7**, 205–23.

WURSTER, D. H. and BENIRSCHKE, K. (1968). Chromosome studies in the super-family *Bovoidea*. *Chromosoma*, **25**, 152–71.

ZARCHI, Y., SIMCHEN, G., HILLEL, J. and SCHAPP, T. (1972). Chiasmata and the breeding system in wild populations of diploid wheats. *Chromosoma*, **38**, 77–94.

ZHDANOV, V. M. (1975). Integration of viral genomes. *Nature*, **256**, 471–3.

Index